AF879178

THE AI REVOLUTION 2050

HOW ARTIFICIAL INTELLIGENCE WILL TRANSFORM HEALTH, ECONOMY AND DAILY LIFE BY MID-CENTURY

MARCIN TKACZYK

SAPMATICS PUBLISHING

Legal and Disclaimer Notice:

Book Cover by Rupa Limbu
Illustrations by Athena Anderson and Maria Hussain
Design Layout by Zakirunissa Karthigeyan

First Edition: October 2024

Paperback: ISBN 978-0-473-72094-0
Hardback: ISBN 978-0-473-72095-7
ePub: ISBN 978-0-473-72096-4
Kindle: ISBN 978-0-473-72097-1
Digital Audiobook: ISBN 978-0-473-72098-8

SAPMATICS PUBLISHING

CONTENTS

"We need to ensure that AI is designed in a way that it aligns with our values and goals, and that we maintain control over these powerful systems."

Stephen Hawking

PREFACE

Why 2050? That may have been the first question you asked when seeing the title. Let me explain.

I did not pick this date at random. Having been in the IT industry for over 20 years, I've experienced many shifts and evolutions within the domain. AI is something different. In a sense, I was raised with AI as it had just started shining on the horizon during my early career, but it was only a weak beam of light initially. The pace of changes and adoption has accelerated dramatically in recent years, making it an ongoing revolution.

Now, after many years, I've come to realize we are somewhere in the midst of a transformation that began in the early 2000s with the resurgence of machine learning, marked by the widespread use of the Internet and other technologies. By 2050, I believe AI will have stabilized and matured significantly. In fact, it may well become a stable force in our lives much sooner. Assuming you and I live to see that time, revisiting this book in 2050 or earlier could offer insights into how accurate—or inaccurate—my predictions and those of experts have proven to be—whether that be something we overlooked as key areas in AI expansion, or some aspect of its expansion abandoned or forgotten about.

Enjoy the reading, and thanks for embarking on this exciting journey through a vision of still mostly unchartered territory with me. We're all part of it.

INTRODUCTION

Introduction

Imagine waking up without an alarm. No incessant blaring, no turning over in bed or hitting snooze to get five more minutes of sleep. In 2050, day begins instead with a gentle nudge from your AI assistant. She's already scanned your schedule, the traffic, weather conditions and set the table for breakfast. You don't have to hurry, as many things are being done for you and the workweek is short. While you get ready, a smart mirror displays the day's headlines tailored to your interests. You step into your self-driving car, which whisks you to your destination while you prepare for meetings. Driving to your workplace is your choice, as virtual reality is enough to work from home, you just enjoy the movement.

Along the way, your AI tunes the car's interior lighting and music to match your mood, enhancing your journey. You catch up on personalized health updates, and yes, you are healthy, as any small imbalances in your body measurements get tracked and immediately taken care of. At lunchtime, you chat with your kids on the transparent screen that unfolds in front, syncing your schedules for afternoon activities. Smart home keeps the fridge fully stocked for when you're back, the walls transform into immersive displays of your favorite landscapes, and a holographic chess partner invites you for a quick game before dinner. All of this has been orchestrated seamlessly by artificial intelligence. You're simply responsible for staying up to date and filling in the gaps.

I've spent a significant part of my IT career working with AI, often without even realizing it (not to mention tools like Microsoft Office's Clippy). When I first started implementing automated decision-making systems, I saw them simply as advanced algorithms, unaware they were actually rising forms of AI. From early days in Poland through working in the global SAP business to my journey in New Zealand since 2010, AI has always been a part of my professional exploration, which only accelerated in the last few years with the rapid advancements in the field. My focus has broadened to encompass Generative AI, NLP, Machine Learning, Computer Vision, and more, actively engaging with global

AI communities and working on projects that seek to integrate these technologies into everyday life, whether it be optimizing traffic flow in the city or personalized health monitoring. My roles have spanned from design to development. But the goal here is not to produce a boring technical textbook. It is about vision.

This book aims to demystify how AI will revolutionize our way of life across multiple domains...

This book aims to demystify how AI will revolutionize our way of life across multiple domains—social, economic, healthcare, educational, and daily activities. It steers clear of unfounded speculation, aiming instead to build upon what we currently understand with regard to AI's capabilities and to thoughtfully predict how they might evolve (which means you might even find certain parts a bit conservative in their outlook). Another goal through this narrative is to foster a deeper understanding of AI's overall structure and components, while making projections about our lives in the not-so-distant future. It's important to keep this real; thus, many examples from everyday life are included.

The content has been structured to cater to a broad audience—from tech enthusiasts and industry professionals, from students to policymakers and general readers curious about future trends. Each chapter delves into different facets of AI, and I hope you will consider it to be in a logical sequence. Reading on, you'll find information about foundational technologies, their implications across various industries, regulatory considerations, and the overarching societal impacts of AI. This book combines currently applied AI domains with insightful projections about their future impact, emerging technologies like quantum computing, and a picture of how the world and AI will look in 2050.

I do not shy away from the controversies of AI, and you may notice that I rather prefer to be more pessimistic than I should be. Though AI has immense potential, there are many risks and ethical considerations. The technology is still relatively new, yet its impact is already immense and continues to grow. These considerations need to be taken seriously. Whether it's enhancing efficiency in businesses or revolutionizing how we manage our health, AI can improve our lives, but it needs to be used in the right way. The opening quote from Stephen Hawking serves as a guiding message that I continue to keep in mind throughout this journey.

At the end you will find a list of 100 predictions that correspond to the chapters and the topics discussed. They are listed in the same order they appear throughout the book, and I've aimed to make them as measurable and concrete as possible. If you'd like, you can visit an interactive version online. You'll find the link in that section. I would love to hear your comments, especially if you think I've missed anything from the big topics. You'll also find an option to build a simple AI project to gain hands-on experience and see how it works in practice.

Finally, the primary idea behind writing these lines was not only to inform but also, with existing knowledge and advances in technology, to prepare us for the inevitable changes that AI is bringing and will continue to bring—transforming our lives in ways not seen since the start of the internet era or industrial revolution. While AI has long been an evolving field, recent years have demonstrated the pace of change and its effects on the world fully deserve to be called a revolution. Today there is not enough said, not enough books written, not enough brainstorming done, and not enough discussions and decision-making efforts surrounding the field.

On the next pages, you will discover how AI works, how it will influence your industry, your health and daily life. You will explore how to identify your best career choices and the optimal path forward. My hope is that this book will provoke thought and encourage you to ask questions. We shouldn't be worried—we have time but we must understand and prepare. The future belongs to those who know and adapt. Stay at the forefront and don't get left behind.

This is the AI revolution 2050.

CHAPTER I

Foundations of AI

Foundations of AI

Despite its prevalence in our daily digital interactions, AI remains a labyrinthine concept to many. However, if you peel back the layers of AI, you can trace its fundamentals to its complex algorithms. Looking into these algorithms more closely can help give you a nuanced understanding of how they work. In this section, we will start explaining those algorithms and explore their unique purposes, as well as the intricacies behind how they work.

Decoding Machine Learning

The Engine of AI

Artificial Intelligence, as defined by the Encyclopedia Britannica, is "the ability of a digital computer or computer-controlled robot to perform tasks commonly associated with intelligent beings" (Copeland, 1998). Such tasks encompass the ability to reason, discover meaning, generalize, and learn from past experiences. At the core of AI's ability to perform these tasks is Machine Learning (ML), a subset of AI that equips machines with the capability to automatically learn and improve from experience without explicit programming.

Machine Learning is distinct from AI in its specificity. While AI is a broader concept concerning machines designed to act intelligently like humans, ML specifically refers to algorithms that allow computers to train on data, learn from it, and make predictions or decisions. ML algorithms can learn in either a supervised or an unsupervised way. Supervised learning involves training a model on a labeled dataset. This means each piece of data has been manually assigned the correct output by a person so that when the AI model is learning, it knows the correct result. As it trains, the model then learns to produce predictions in line with these results based on any similar input data provided. Unsupervised learning, in contrast, uses no labels. Instead, it identifies patterns and relationships in the input data (GeeksforGeeks, 2024).

The recommendation systems used by streaming services like Netflix and Spotify are some great examples of machine learning in action. These systems analyze your past behavior and compare it with that of others to recommend movies or songs you might like. ML algorithms are employed in a variety of other fields, though. For example, in finance, they are used for predictive analytics to assess loan risks or detect fraudulent transactions.

The quantity and quality of the data an ML model is trained on directly influences the accuracy of the model.

The training of ML models is a critical process in any of its applications. This is because the quantity and quality of the data an ML model is trained on directly influences the accuracy of the model. To work well, ML models need to be trained on large sets of data. These datasets can be hard to build or acquire, but once they exist, they enable ML systems to make informed predictions and decisions that can revolutionize an industry. A well-trained ML model can sometimes provide insights that are not even apparent to human analysts.

However, machine learning is not without its limitations and difficulties. One of the major issues is data bias—if the training data is biased, the model's predictions will be biased, leading to unfair or incorrect outcomes. Another challenge is overfitting. That's because a model learns the details and noise in the training data to the extent it overshadows the relevant information on which the model should be trained. This can then negatively impact the performance of the model on new data.

The Evolution of AI

A Historical Perspective

By studying the evolution of AI, from its nascent stages in the mid-20th century to the sophisticated systems we interact with today, we can gather insights into the improvements and complications the technology may present us with in the future. That evolutionary journey began at the Dartmouth Conference in 1956, where a small group of leading thinkers in AI, computer science, and IT coined the term "artificial intelligence"(Solomonoff, 2023). This ignited the first wave of optimism and investment in AI, leading to pioneering programs like ELIZA and Perceptron, which demonstrated machines could perform tasks previously reserved for humans, such as language translation and simple decision-making.

Unfortunately, there were major technical challenges. The first AI Winter occurred between 1974 and 1980, a period characterized by reduced funding and waning interest in AI research. The limitations of technology at the time, namely insufficient computational power to process complex algorithms, stymied further development.

In the 1980s, things changed with advances in machine learning and the development of expert systems, which are designed to mimic the decision-making abilities of a human expert. These advances enabled AI to overcome some of the issues that had plagued it earlier, yet it still proved to have too many limitations for widespread use. High costs, the limited commercial success of AI systems, and criticism from prominent figures in the field led to a considerable reduction in interest and investment once again. This culminated in a second AI Winter from 1987 to 1993 (History of Data Science, 2021).

In the early 2000s, the field of machine learning saw a major resurgence, driven by advances in computing power and the availability of large datasets. This period marked the transition from traditional AI approaches to more data-driven methods, enabling more sophisticated algorithms and models. Techniques such as support vector machines, decision trees, and early neural networks began to show practical utility across various industries, laying the foundation for future breakthroughs in deep learning. Renewed focus on machine learning helped propel AI from theoretical research into practical applications, ultimately leading to the advanced systems we see today.

THE EVOLUTION OF AI: A HISTORICAL PERSPECTIVE

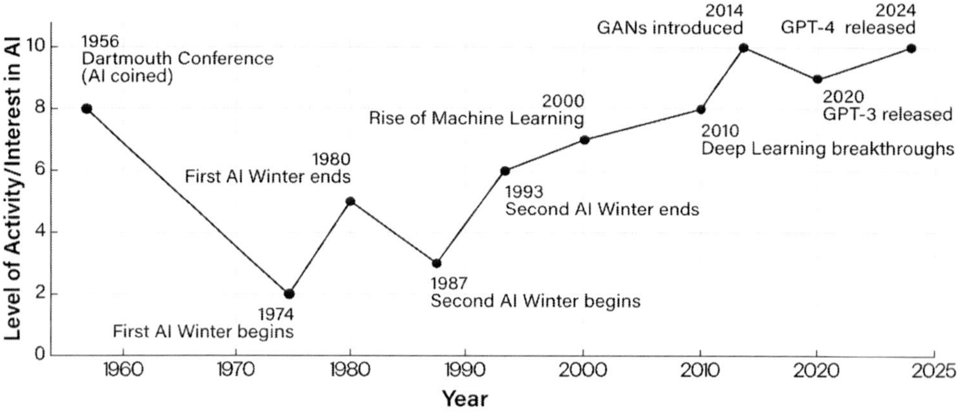

Several key figures have played pivotal roles in sustaining the momentum of AI research. In 1950, Alan Turing, often hailed as the father of theoretical computer science and artificial intelligence - and whom you can see in the opening chapter page - proposed the idea of a test that could determine whether a machine could "think." This test, known as the Turing Test, remained a fundamental part of the AI discourse through each AI winter and even into the present day, providing an ideal for AI researchers and scientists to aim for as they strive to create more powerful systems (The Editors of Encyclopedia Brittanica, 1998).

Another significant figure in the field of AI, Geoffrey Hinton, is known for his revolutionary work on a subset of AI called deep learning. Most of his work took place after the second AI winter, and it brought with it new ways of thinking about AI systems. Instead of the simpler, guided learning methods ML uses, deep learning utilizes larger and more complex datasets to create a system of neural networks that mimic how the human brain works. Hinton is often called "the godfather of AI" for his groundbreaking work, which revolutionized the capabilities of AI systems in pattern recognition and predictive analytics.

There are numerous technological breakthroughs that have shaped modern AI. In 1986, backpropagation, an algorithm first invented in the 1970s but not widely appreciated until later, became a cornerstone used for training neural networks. Nowadays, backpropagation is critical for enabling AI to learn from complex data sets and improve over time (McGonagle, Shaikouski, & Williams, n.d.).

Another technology, Generative Adversarial Networks (GANs), has provided more recent improvements. GANs were introduced by Ian Goodfellow and his colleagues in 2014. They represented a novel approach to generative modeling, pitting two neural networks against each other to generate new data that is indistinguishable from real data. These technologies have both enhanced the performance of AI and expanded its applications, as they enable systems to use more data (Giles, 2018).

In the present day, AI's integration into industries such as healthcare, finance, and manufacturing has revolutionized operational processes and customer service in a myriad of ways. We'll explore more about those changes in the coming sections.

To mention one example, you may have heard of ChatGPT by OpenAI. ChatGPT has recently demonstrated substantial advancements in natural language processing and conversational AI. It can generate human-like text based on a given prompt, which enables it to be used for all sorts of things, from customer service to content creation (Pegasus One, n.d.). Yet, a surge of interest in ChatGPT has raised important questions about the role of authenticity in digital communication and the potential for ChatGPT to be misused, such as for the creation of "deep fake" content or the spread of misinformation.

AI's continued evolution is inevitable, so understanding where it has been can help us prepare for the future. Its trajectory so far has been anything but linear—there have been advancements, setbacks, recoveries, and doubt. Each phase has set the stage for the next one, and the same will probably remain true as we peer toward a future where its integration into society will one day be seamless and transformative.

Weak AI (ANI) vs. Strong AI (AGI)

What AI Can and Cannot Do

Before we proceed, it's important that we clarify and can distinguish between Weak AI, also known as Narrow AI, and Strong AI, or in other words General AI.

Weak AI is the type of AI that is currently in widespread use. It's designed to handle specific tasks such as language translation, facial recognition, or driving an autonomous vehicle. Weak AI systems operate under a narrow set of constraints and limitations, excelling in particular domains but lacking the broader cognitive abilities that define human intelligence.

On the other hand, Strong AI remains a theoretical and aspirational goal. That goal embodies the capability of machines to understand, learn, and apply knowledge across a diverse range of tasks, mirroring human cognitive abilities (Northwood Tech, n.d.).

In the present day, Weak AI has transformed industries by automating complex tasks that were previously thought to be the exclusive domain of humans. For instance, in healthcare, as we'll see in later chapters, AI algorithms can analyze medical images with a precision that matches or even surpasses human radiologists. In manufacturing, AI can make predictions about demand to reduce the incidents of too much or too little stock. However, in all these cases, Weak AI is fundamentally limited by the absence of consciousness and genuine understanding. It operates through predefined algorithms and data patterns, with a very specific and narrow-minded goal, lacking the ability to empathize or grasp nuances intrinsic to human interactions.

Weak AI has significant limitations when it comes to creative or intuitive thinking. A machine learning model trained to generate music can produce new compositions by analyzing vast datasets of existing music, without an honest understanding of creativity or emotional resonance (Pal, 2024). Without any real comprehension of the art it is engaging with, this leads to results that lack what is necessary to really draw a

Aspect	Weak AI (ANI)	Strong AI (AGI)
Definition	Narrow AI designed for a specific task or a set of related tasks	General AI with human-like cognitive abilities and understanding across a wide range of tasks
Scope	Limited to specific applications (e.g., chess playing, language translation)	Capable of performing any intellectual task that a human can
Capabilities	Performs predefined tasks with high efficiency; does not understand context beyond its programming	Understands, learns, and applies knowledge in various contexts; capable of reasoning and problem-solving
Examples	Siri, Alexa, Chatbots, Recommendation systems, Image recognition systems	Hypothetical future AI (e.g., AI depicted in science fiction, such as HAL 9000, Data from Star Trek)
Learning	Uses machine learning algorithms to improve performance in specific tasks	Possesses the ability to learn and adapt in a human-like manner across a wide array of activities
Consciousness	No self-awareness or consciousness; operates purely on input-output basis	Hypothetical self-awareness and consciousness, akin to human consciousness
Understanding	Lacks understanding of context, nuance, and deeper meaning	Understands and processes context, nuance, and deeper meanings like a human
Decision-Making	Limited to decision-making within its narrow task; relies on preprogrammed algorithms	Capable of autonomous decision-making in diverse and unforeseen scenarios
Flexibility	Rigid and task-specific; unable to generalize knowledge beyond its programmed domain	Highly flexible; able to generalize knowledge and apply it to new and diverse problems
Dependence	Dependent on human input for learning and updates	Autonomous in learning and adapting to new situations
Ethical Concerns	Bias in data, lack of transparency, limited accountability	Potential existential risks, ethical decision-making, and alignment with human values
Current Status	Widely implemented and used across various industries	Theoretical and speculative; not yet achieved

listener in. Similarly, while AI has made impressive strides in natural language processing, these systems often falter when faced with the complexity and subtlety of human language. AI normally requires language to be precise and clearly structured. In real life, it's often not like this. The nuance of sarcasm, irony, or cultural context frequently escapes the grasp of AI, leading to misunderstandings or overly literal interpretations (Gillis, n.d.).

How do we imbue machines with wisdom?

Weak AI can struggle to deal with unforeseen circumstances. When training on data, AI operates in a structured environment where parameters are well-defined. This translates well to situations like playing a game of chess or navigating from point A to point B. However, when confronted with a situation requiring out-of-the-box thinking or ethical judgment, AI is more likely to falter, lacking the ability to reason abstractly or consider the broader implications of their actions beyond what they've been trained on. For example, a weak AI system trained to help with inventory might not know what to do if a shipping issue arises or there is an unforeseen manufacturing defect.

While some AI systems can simulate basic forms of social interaction and even mimic human conversational styles, they do not form genuine emotional connections or possess true empathy. It can be a problem when the emotional nuances of a conversation demand an understanding of the social and cultural contexts that inform the emotions a person is experiencing. For instance, AI-driven chatbots can offer customer service by responding to queries with programmed empathy and politeness, but these interactions lack the depth and understanding needed to discern why someone is crying or whether someone is masking an emotion they might be feeling (Agarwal, 2022).

Strong AI, on the other hand, is an ideal that doesn't have these limitations. Achieving an AI that genuinely mirrors human intelligence, though, will involve many technological advancements, along with deep philosophical and ethical considerations. How do we imbue machines with wisdom? Can a machine ever truly understand human values or ethical principles? These are profound questions we continue to grapple with right now. We will revisit this topic in the closing chapter as we explore potential future scenarios of AGI.

Deep Learning Demystified

Understanding Neural Networks

Deep learning is a subset of machine learning that has been instrumental in advancing the capabilities of artificial intelligence. At the center of this advancement are structures called neural networks—inspired by how the human brain works and designed to recognize patterns and make decisions (Awan, 2023). It might sound like we're getting awfully close to strong AI, but the reality is neural networks still have quite a ways to go. We haven't unlocked all the secrets about how the brain works... yet.

However, we have made some extraordinary strides in that direction, thanks to Geoffrey Hinton, the "godfather of AI" we mentioned earlier, and others who worked alongside and after him. At its core, a neural network is an algorithm intended to recognize underlying relationships in a set of data. Modern neural networks do this with layers of nodes, or "neurons," each designed to perform specific yet interconnected tasks (Awan, 2023).

The architecture of a neural network, often referred to as its structure, empowers it to perform complex functions such as image and speech recognition. Each neuron in the network processes a small amount of task-specific information before passing it on to the next layer of neurons. Consequently, the network is able to transform input data into outputs through a process known as forward propagation (Machine Learning in Plain English, 2023). Meanwhile, backpropagation, which we mentioned earlier, refines the accuracy of the output by adjusting the weights of connections. This might sound complex. The essence is that forward propagation is the process by which a neural network does stuff, and backward propagation is the process by which it learns from its mistakes, enhancing its performance over time. Furthermore, neural network layers can be categorized into three types: the input layer, which receives the raw data; hidden layers, which analyze the data; and the output layer, which delivers the final decision or prediction (Machine Learning in Plain English, 2023).

Deep learning is still expanding and growing as a viable field of application for AI. Deep learning models can already identify and categorize objects in images with accuracy rates that improve upon human performance (NYU, 2021), but there are also other things it can't do yet. At present, its applications are practical in some of the following areas:

- **Medical imaging:** It helps diagnose diseases with X-rays and MRIs with a high degree of accuracy.
- **Speech recognition:** Deep learning has been the driving force behind the development of voice-activated assistants like Siri and Alexa. These systems rely on neural networks to understand and process human speech, enabling them to comprehend and respond to voice commands.
- **Natural language processing (NLP):** Deep learning models are used to translate text, summarize articles, and even generate human-like written content.

As it evolves and algorithms become smarter, deep learning has the potential to revolutionize various sectors of the economy and aspects of our daily lives. It is conceivable that deep learning models will become even more integrated into industries, further automating tasks and providing new insights from large quantities of data that get us ever closer to Smart AI.

We are still in the early stages of AI. As we continue to explore and innovate, the boundaries of what deep learning can achieve will extend further. The only questions are how far and how fast.

From Theory to Application

How AI Learns

From basic tasks to complex decision-making processes that increasingly mimic human cognitive functions, AI can already be utilized in a plethora of ways across many industries. At the heart of these applications are the fundamental learning algorithms of AI, which include decision trees, clustering, and reinforcement learning. Each of these play a pivotal role in how AI systems interpret and interact with the world around them.

Decision trees are a type of supervised learning algorithm that model decisions and their possible consequences with a system somewhat reminiscent of a flowchart. These are called trees because each part of the tree or "node" demands a decision that will cause the tree to branch off into other nodes. Eventually, these decisions will end at what's called a "leaf node," where there is no decision left to make. In a decision tree, each internal node (a node that is not a leaf node or the root note) represents a "test" on an attribute, each branch represents the outcome of the test, and each leaf node represents a class label, which can then be applied in a transparent and easily understandable way (IBM, n.d. b).

Clustering, a form of unsupervised learning, groups a set of objects in such a way that objects in the same group are more similar to each other than to those in other groups. This technique is widely used in market segmentation, organizing computer resources, social network analysis, and astronomical data analysis (Ali, 2022).

Meanwhile, reinforcement learning is a type of dynamic programming that trains algorithms using a system of rewards and penalties. It's an unsupervised learning method that teaches the model to make a sequence of decisions by trying out new strategies to maximize a reward metric. This can be done without human intervention and with minimal prior knowledge of the system (Bajaj, 2024).

Many more modeling techniques exist, and they vary broadly in scope. They range from statistical models that predict probabilities based on input data to machine learning models that can learn and make decisions

AI LEARNING PROCESS

The AI learning process from data collection to model training and deployment

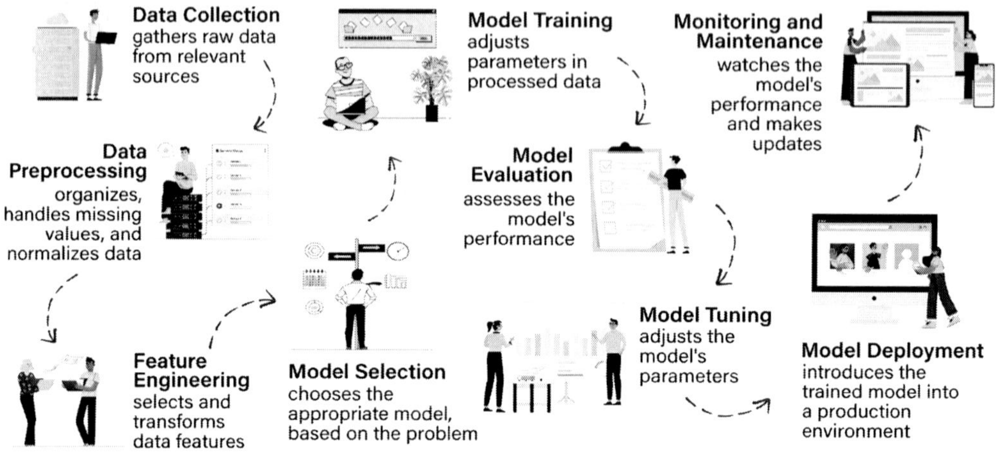

from data and hybrid models combining both elements. These models define the approach AI systems take to process and learn from data. For example, Tesla's Autopilot system utilizes a form of continuous learning where the vehicle constantly collects data from its environment and sends this information back to Tesla (Harris, 2022). The data is then used to improve system functionalities and make the self-driving algorithm more accurate for all vehicles.

That said, as you may realize by now, the efficacy of these algorithms depends heavily on the quality and diversity of data they are trained on. That's why big data plays such an integral role in AI development, and it's one reason why AI struggled to get off the ground at first, suffering through two winters where funding and interest in it waned. Without the ability to train on large sets of quality data, AI is more likely to produce erroneous results.

In this context, one challenge AI faces is the issue of data privacy. Laws like the GDPR in Europe have put stringent measures in place to measure how data is collected, stored, and used. Additionally, another challenge it faces is around security concerns. Because of all their data, AI systems are often targets for cyber-attacks that could potentially lead to data being leaked to malicious actors. Furthermore, the computational cost of training AI models remains high and can present a challenge in scaling AI solutions, necessitating ongoing advances in computing power and efficiency.

In summary, the future of AI learning is set to be revolutionized by further integration with big data, propelled by advances in data storage and processing technologies. As cloud storage becomes more secure and processing power continues to grow, models will become more sophisticated, with enhanced capabilities to process and learn from data in real time. This will open new avenues for AI applications across different sectors, possibly transforming how we interact with technology on a fundamental level.

AI's ability to learn and adapt is what truly sets it apart from traditional computational models. This adaptability will only become more pronounced in the future.

AI Development Frameworks
Catalysts for Innovation

Without getting too technical, I find it beneficial to introduce the most popular tools for AI development, how they are currently being used, what they can, and what their future impact may be. We will be referencing them later as we continue. The current landscape of artificial intelligence is built upon a foundation of robust development frameworks that help to streamline the creation of AI models through powerful and innovative platforms that are accessible to developers around the globe. Among these, TensorFlow, PyTorch, and Scikit-learn stand out as some of the most influential, each contributing uniquely to the AI ecosystem.

TensorFlow, developed by Google, offers a comprehensive suite of tools, libraries, and resources that allow for the building and deploying of machine learning models at scale. Known for its flexibility and extensive capabilities, TensorFlow supports both research and production, making it a preferred choice for developers aiming to push the bounds of AI technology.

Similarly, PyTorch, created by Facebook's AI Research lab, has gained popularity for its user-friendly interface and dynamic computational graph. These things allow for changes to be made on the fly during execution. This flexibility is particularly valuable in the research phase of AI development, where iterative experimentation is crucial. PyTorch is celebrated for its speed and efficiency in model experimentation and prototyping, which accelerates the development process.

Scikit-learn, on the other hand, is revered for its simplicity and efficiency, especially suitable for those new to machine learning. It provides a range of supervised and unsupervised learning algorithms via a consistent interface in Python. It's good for educational purposes and smaller-scale projects, as these situations require a straightforward approach without the need for the extensive computational resources frameworks like TensorFlow might necessitate.

These frameworks and others like them serve as platforms that accelerate the entire cycle of research and development. They allow new developers to build off what's already been created rather than starting from scratch. This acceleration can be a big boon when dealing with tight deadlines and a rapidly evolving industry such as AI. For instance, TensorFlow's extensive libraries contain the ability to process, analyze, and infer from data at an unprecedented scale. Projects that utilize TensorFlow, whether in the area of consumer applications, complex scientific research, or something else, can thus take advantage of these libraries to enhance their results and achieve those results much quicker.

Such platforms play a vital role in fostering collaboration within the AI community. Many are open-source by nature, which means they encourage developers from around the world to contribute to the ecosystem, whether by refining the frameworks themselves or by sharing implementations of novel algorithms that push the boundaries of what AI can achieve. This collaborative environment can accelerate advancements in AI technologies and democratize access to the frameworks.

There are many examples of AI development frameworks contributing to applications across various sectors. For example, TensorFlow has been used for things such as fraud detection at PayPal, personalized recommendations on Spotify, and the identification of the anatomy of MRIs on the brain at GE Healthcare (TensorFlow, n.d.). Similarly, PyTorch has contributed to many applications in the area of natural language processing (NLP) and has been used by Amazon Advertising to assist in flagging ads that are non-compliant with Amazon's terms (Kanungo, 2022).

The future of AI development frameworks looks even more promising as they begin to integrate with emerging technologies like quantum computing and the Internet of Things (IoT). The synergy of these is expected to unlock new advancements in our society, such as faster processing speeds and enhanced applications with IoT devices, leading to smart systems that can do more and be created more easily.

It is expected development frameworks will continue to play a critical role in catalyzing innovation.

Coding assistants like GitHub Copilot, TabNine, and Kite are some promising and potentially revolutionary applications of AI. These tools leverage machine learning models to provide real-time assistance to developers by suggesting code completions, identifying bugs, and optimizing code performance. This can both boost productivity and enhance the quality of code, making the development process more efficient and less prone to errors.

It is expected development frameworks will continue to play a critical role in catalyzing innovation. They should evolve at an accelerated pace, enabling the integration of AI with a wide swath of different technologies so that it touches every aspect of society. From their role promoting the collaboration of engineers across the world to their sophisticated algorithms that enable customer-facing solutions to be created more quickly, they have enduring potential.

The Role of Quantum Computing in Advancing AI

Quantum computing represents a seismic shift in our approach to data processing. It's poised to revolutionize a myriad of fields, including artificial intelligence. Unlike classical computing, which relies on bits as the smallest unit of data represented in one of two states, 0 or 1, quantum computing uses quantum bits, or qubits, which can exist in multiple states simultaneously thanks to the principles of superposition and entanglement. This allows quantum computers to process vast amounts of data at speeds unattainable by their classical counterparts.

AI and quantum computing can offer a lot when it comes to working together on algorithm design. AI can craft algorithms that optimize the use of quantum states, thereby enhancing the efficiency of quantum computations. Quantum algorithms can be designed to accelerate machine learning processes by performing computations in ways that are fundamentally different from classical computers. This capability could potentially reduce the time required for data training and model development from weeks to mere hours, speeding up the iteration cycles in AI research and application.

Similarly, quantum computing has the potential to disrupt the field of cryptography. AI-enhanced quantum computing could create new forms of encryption far more secure than anything currently available, safeguarding data against even the most sophisticated cyber threats. Conversely, the same technology could be used to break much of the encryption that currently protects our digital communications, so it's essential we be prepared and evolve quickly.

The capabilities of quantum computing extend into other fields requiring complex computation, too, such as drug discovery and climate modeling. In drug discovery, quantum computing might analyze molecular interactions at a level of detail impossible for classical computers, accelerating the development of new medications. Similarly, in climate science, quantum computers could model complex climate systems with new levels of precision, providing more accurate forecasts and enhancing our understanding of climate change impacts

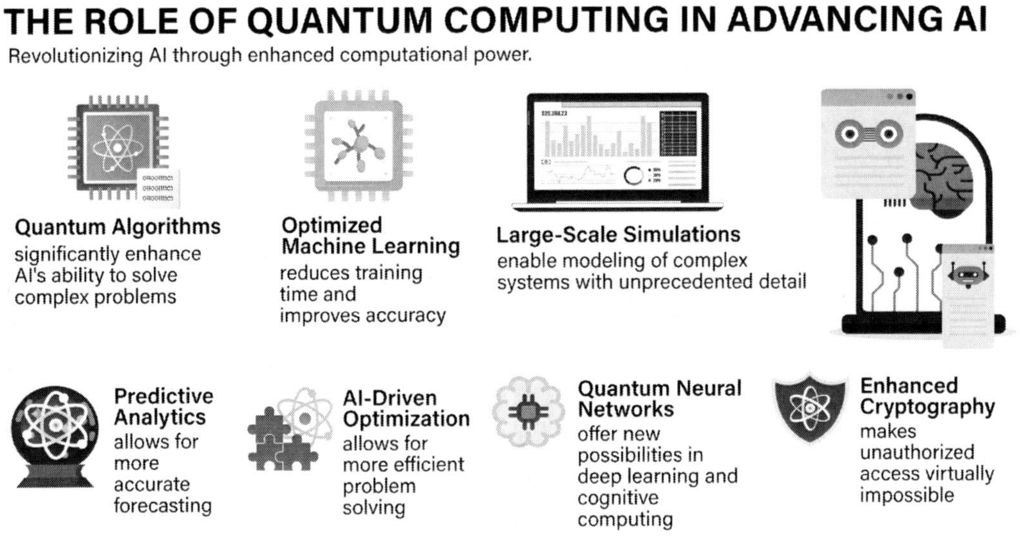

In addition to its possible dangers in the field of cryptography, quantum computing has several other challenges. Current quantum computers are highly sensitive to external disturbances, which can cause errors in calculations, a phenomenon known as quantum decoherence. Additionally, the physical hardware required to build quantum computers is both expensive and technically complex, limiting accessibility to only those with knowledge and resources. However, ongoing research and development should eventually lead to improvements in these areas. Many current research projects are focused on qubit coherence times, error rates, and scalability

Recently, Google's Quantum AI Lab has already started to make some strides in the right direction with its Sycamore processor. In 2019, they announced Sycamore had achieved quantum supremacy by performing a specific task in 200 seconds that would take the world's most powerful supercomputer thousands of years to complete (Martinis & Boixo, 2019). This landmark achievement demonstrated the potential of quantum computing to tackle problems beyond the reach of classical computers while also highlighting the future possibilities of quantum technology in enhancing AI capabilities. The Google lab continues to explore avenues to integrate quantum computing with AI, focusing on solving complex optimization problems, improving machine learning algorithms, and simulating intricate molecular structures that could have far-reaching implications across multiple scientific domains.

The future of AI and quantum computing looks bright. These two fields will work together, leading to advancements that can disrupt industries and classical methods far and wide, from cryptography to climate science and even the algorithms that dictate how we interact with our technology. We will revisit this breakthrough technology in the concluding parts of our exploration.

Anticipating the Future

A Journey from Now to 2050

AI development is not just about technological advancement. It's about understanding the ethical, societal, and economic impacts these technologies engender. Concerns have only grown louder in recent years, and I expect them to remain key issues that guide our way forward. As the year 2050 grows nearer, it will be more important than ever to make sure our systems use responsible AI—technologies that prioritize ethical considerations, data privacy, and inclusivity.

One of the most crucial lessons we have learned so far is the significance of interdisciplinary collaboration in AI research and development. The integration of insights from fields such as psychology, ethics, and social sciences, alongside computing and engineering, is necessary to develop AI systems that are effective, socially responsible, and ethically sound. After all, if AI hopes to provide value in these fields, it needs to be able to learn from experts who can share their insights. As we venture into developing more advanced AI systems, including efforts toward achieving Strong AI, these things will not change. Any good AI system will require a deep understanding of human cognitive processes and ethical reasoning from solid teachers and knowledgeable sources.

It is likely the evolution of AI will not proceed without wrinkles. In the past, the periods of stagnation have tested its resilience, spurred by waning interest and investment in AI research. These periods highlight the industry's vulnerability to hype cycles and unrealistic expectations. They are reminders of the importance of setting realistic goals, transparently communicating the capabilities and limitations of AI technologies, and maintaining steady research funding to ensure sustained progress regardless of market sentiments.

AI is well on its way to reshape multiple industries. Healthcare, transportation, climate change, and cryptography are some of the areas where it could be most effective. If we remain focused on what is realistically achievable, it's possible we can see steady results. This means things such as personalized medicine tailored to each individual's genetic makeup in industries such as healthcare and perhaps autonomous vehicle driving systems in transportation. Meanwhile, realistic goals in tackling climate change might revolve around using AI's ability to analyze large datasets quickly and accurately to develop more effective strategies for energy use, conservation efforts, and disaster response

Additionally, it will become even more important for the public to become educated and engaged about the ways in which AI works. Only by increasing AI literacy and involving diverse communities in the development and governance of AI can we develop these technologies in ways that are aligned with public values and societal needs.

Ultimately, ethical issues might become a key obstacle to achieving further progress. We must not forget the AI setbacks that came before, keeping our expectations in check so as not to overhype the technology. If we can push through these roadblocks, changes will come. AI can revolutionize everything from healthcare to transportation and even our modern IoT devices by 2050. The journey ahead will require careful navigation, but these potential applications point to a transformative effect if we can get it right.

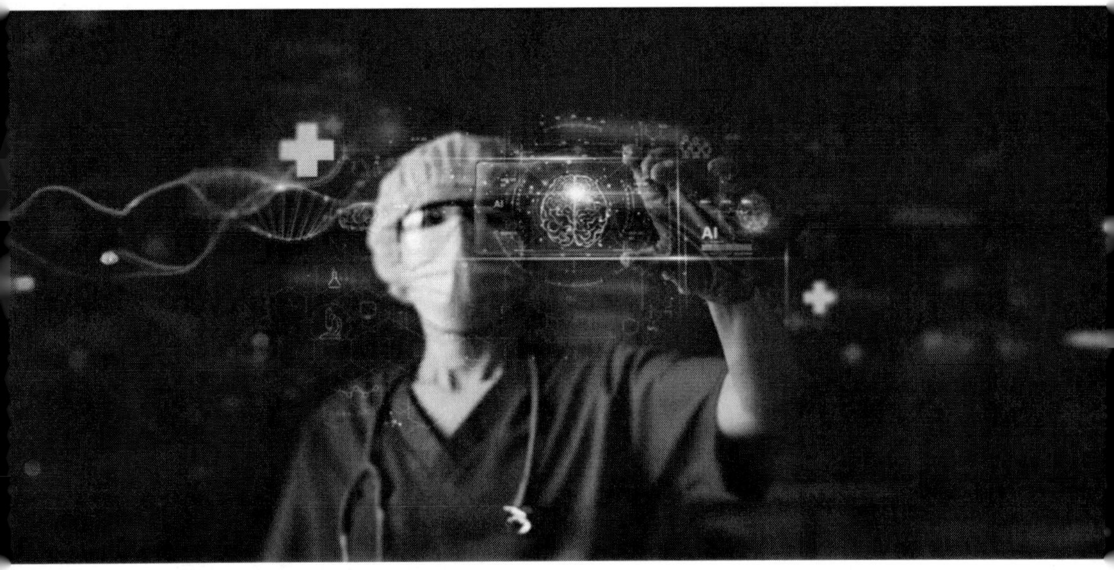

CHAPTER 2

AI in Industry

AI in Industry

AI HAS APPLICATIONS across all manner of industries. From the healthcare sector to retail, manufacturing, and even the defense sector, we will explore these applications, forecasting both what may come in the future and what's already happening that you might not know about.

Healthcare

Revolutionizing Patient Care

In healthcare, AI's prowess is already becoming evident by enhancing diagnostic systems. Through advanced imaging and pattern recognition capabilities, AI has made big strides in improving the processes in radiology, pathology, and several other fields that rely heavily on accurate diagnostics. Complex algorithms trained on vast datasets can now identify subtle patterns in imaging data invisible to the human eye, such as in the case of a 2022 study where researchers at Mount Sinai discovered that microscopic abnormalities could be used to detect the onset of Alzheimer's (Mount Sinai, 2022). In another research study from 2019, AI was shown to have beneficial effects for detecting breast cancer. NYU Langone's Department of Radiology showed AI improves radiologists' ability to correctly diagnose breast cancer by 37% while requiring 27% fewer biopsies to confirm suspected tumors (NYU, 2021). The precision of AI in these applications alleviates the workload on medical professionals, allowing them to focus more on patient care rather than administrative tasks.

Personalized medicine is one of the most revolutionary ideas brought forth by AI. Leveraging data from an individual's genetic profile, AI algorithms could predict the most effective treatment plans tailored specifically to each patient. This approach would be a significant shift from the traditional "one-size-fits-all" methodology, promising treatments with higher efficacy and minimal side effects. Personalized medicine can be particularly effective in the treatment of chronic diseases such as diabetes and cancer, where the interplay of genetic factors influences the effectiveness of various treatments. By analyzing genetic information in conjunction with medical records and even lifestyle data, AI can base its recommendations on a holistic view of a patient's health, enabling bespoke treatment strategies that are both proactive and preventive. There is still a lot of work to do before the data for this sort of thing can be compiled and made available for AI systems. However, it's conceivable we may get there by 2050.

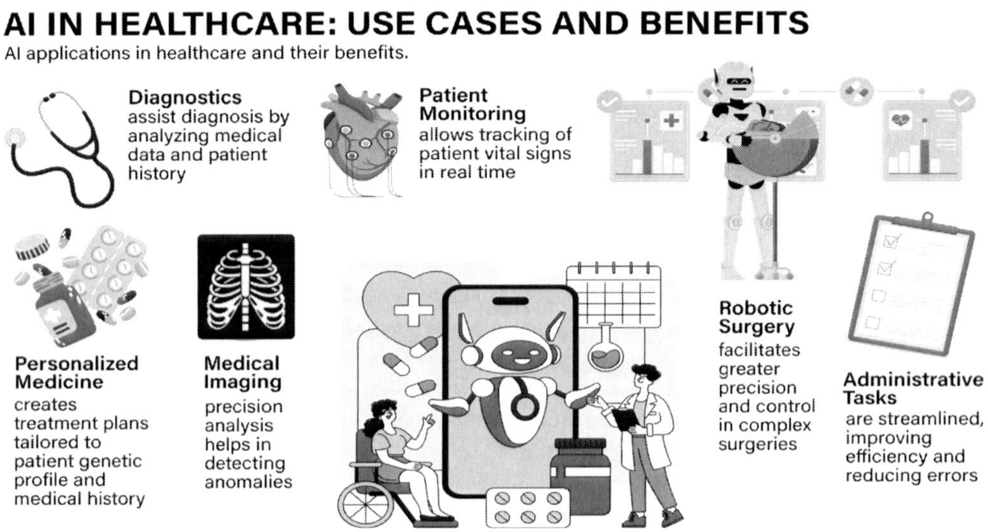

The integration of robotics and AI is already redefining surgical procedures. AI-driven robotic systems enhance the precision and control of surgical interventions, making complex surgeries less invasive and reducing recovery times. These robotic assistants perform precise surgical tasks with consistency and accuracy beyond human capabilities. In 2022, an enhanced version of a Smart Tissue Autonomous Robot (STAR) designed by a team at Johns Hopkins University performed a successful laparoscopic surgery on a pig without any guidance from a human hand (Graham, 2022). Axel Krieger, an assistant professor at Johns Hopkins, said that the AI and robot combination "produced significantly better results than humans performing the same procedure." AI and robotics have also been combined to make meaningful strides in procedures like knee replacements (Joint Preservation Institute, n.d.)

and even offer promising prospects for complex procedures such as delicate neurosurgeries.

Outside the confines of hospitals, clinics, and research laboratories, AI-powered virtual health assistants show increasing promise in their bid to become an integral part of everyday health management (Teo, 2022). These assistants provide health monitoring and support, offer advice, remind patients about medications, and even manage chronic conditions without requiring that the patient leave their home. They're designed to ask simple questions, then analyze the input and give suggestions based on observations from both training data and their ongoing experiences. Their ability to learn and adapt to the user's needs over time makes them an invaluable companion for health management. This technology is particularly beneficial for elderly patients or those with chronic conditions who require continual monitoring, thereby bridging the gap between periodic doctor visits and a continuum of care that is both responsive and preemptive.

Smart health devices are showing AI can be beneficial for health monitoring and preventative care as well. Technologies like the Apple Watch can track heart rate and other health indicators, while technology like Sleep Number's SleepIQ, built into their beds, can detect sleep time, restful and restless sleep, average heart rate, and respiratory rate. Combined with AI, this data can be used to provide users with actionable insights about their health, encouraging proactive health management and reducing the need for frequent in-person consultations with doctors.

Along similar lines, these sorts of shifts could do a lot to improve predictive healthcare. By analyzing patterns and correlations in vast amounts of health data, AI models could forecast health trajectories based on patient history, family history, and data from IoT devices and then use the information to identify risk factors early on. This predictive capability would allow for interventions that are more targeted and effective than what we have now, preventing diseases from developing or progressing. For instance, AI systems could analyze data from various sources, including electronic health records, genetic data, and lifestyle information, to determine an individual's risk of developing heart disease, cancer, or diabetes. This has the added benefit of reducing the overall burden on healthcare systems by curtailing the incidence of preventable diseases.

You should already be starting to see where healthcare could go by 2050, but let me present yet another application. The integration of AI with biotechnology, particularly in the fields of gene editing and synthetic biology, presents yet more potential for the budding field. AI's ability to process and analyze complex genetic data at scale could enhance our understanding of genetic disorders and the impact of genetic variations on health. Through its ability to identify even microscopic abnormalities, like in the case of the study we mentioned about Alzheimer's earlier, this could enable the development of gene therapies that precisely target and modify genes responsible for certain diseases, leading to cures for conditions that were previously considered untreatable. As these technologies advance, it's conceivable AI will bring us some degree closer to an ideal reality where genetic disorders are no longer a life sentence but a curable condition.

Automotive

Beyond Autonomous Vehicles

AI's influence on the automotive industry extends beyond autonomous vehicles. It is poised to redefine every aspect of automotive technology, from predictive maintenance to enhanced driver experiences and smart traffic management. By mid-century, we could see this evolution as something that yields heightened efficiency and safety, and a paradigm shift in how we perceive and interact with vehicles.

In automotive production lines, AI-driven robots and systems have the capacity to streamline operations, predict demand, and cut manufacturing costs (Intellias, 2020). These intelligent systems can quickly adapt to new vehicle models or modifications in production processes, reducing the time and cost associated with retooling. Applications like Computer Vision can identify anomalies with visual and diagnostic sensors so these issues can be fixed before they cause serious quality issues. These systems can operate faster and at much higher accuracy than a human inspector.

Predictive maintenance is a prime example of how AI contributes to the automotive industry. Traditional vehicle maintenance schedules are based on mileage or time intervals, but AI introduces a more dynamic approach. Through its integration with IoT devices such as sensors on the car, AI systems can continuously monitor the vehicle data, combine what is noticed with information learned from historical maintenance records, and then predict when a vehicle part is likely to fail or when maintenance is necessary. These predictions are, of course, only as powerful as the data from sensors and historical data that exists out there. However, when AI technology advances, it's conceivable these datasets will get bigger and more accurate, due to more refined and/or numerous sensors to monitor key features on your car.

It is poised to redefine every aspect of autonomous technology...

The in-car experience is an area where AI is making significant inroads already now. Modern vehicles are equipped with a myriad of sensors and data inputs that AI systems can analyze to customize the driving experience. From adjusting the internal climate to selecting music based on the mood and preferences of the occupants, AI can transform a vehicle in a personalized and smart way. In-car AI systems are already being developed to improve safety on the roads. In 2019, Bosch developed an interior monitoring system that uses cameras and AI to detect when a driver is distracted or tired. Their technology then warns the driver, recommends they take a break, or even slows the car down (Frangoul, 2019). Furthermore, AI enables adaptive interfaces to learn from driver interactions, improving over time to become more intuitive and responsive. For example, voice recognition technology powered by AI, such as what's used in Amazon's Alexa device, can understand natural speech patterns and accents. This has the potential to one day allow drivers and passengers to interact with vehicle systems through simple voice commands, such as by turning the radio on/off or adjusting the defogger without the need to look for the button or the right setting on the dial.

In 2050, we can expect to see the beginning of smart cities that use AI for traffic management as well. In fact, I was already part of such a conceptual project. AI algorithms can analyze traffic data in real time to optimize traffic flow and reduce congestion. These technologies are already being integrated in a variety of cities around the world, such as in London, where a recent smart roadmap introduced goals for smart street infrastructure and better data gathering (Stofan, 2023). In the future, these systems could adjust traffic light sequences during peak times or in response to traffic jams or accidents, improving overall traffic efficiency.

> ***... continuing refinements in traffic management can be better for the environment, leading to reduced vehicle emissions and better air quality.***

Furthermore, AI-driven navigation systems are already revolutionizing the way we travel from one destination to the next. Apps like Google Maps and Waze use real-time traffic updates to offer alternative route suggestions, further easing congestion and lessening the time drivers need to spend on the road. In the long run, continuing refinements in traffic management can be better for the environment, leading to reduced vehicle emissions and better air quality.

AI integration into the automotive industry is set to redefine the boundaries of what vehicles and our traffic systems can achieve. From making cars safer and more enjoyable to drive to transforming how they are produced and maintained, AI's role in the automotive sector is just beginning to take shape. As we look toward a future dominated by smart technology, AI will become more integrated into the automotive industry in all the aforementioned areas, propelling us forward into a future that is more climate-friendly and more efficient.

Retail

Customizing Consumer Experience

AI's role in revolutionizing retail extends from the back end of inventory management to the front lines of customer interaction. In the area of inventory management, predictive analytics gives retailers the ability to forecast product demand with newfound accuracy. These systems are capable of accounting for seasonal changes in demand and can incorporate other data, like past sales data, real-time shifts in consumer behavior, emerging trends, and more, to make highly accurate predictions. This foresight allows for systems to adapt to changes much faster than if humans had to notice the shifts in consumer sentiment and adapt, thereby minimizing the occurrences of overstock and understock scenarios, and reducing costs and maximizing revenue.

AI-powered recommendations can be used to improve the retail industry as well. These recommendations personalize the shopping experience by analyzing a wealth of data, including past purchases, browsing history, and consumer preferences, and then using this data to suggest products that align closely with the customers' tastes and needs. For example, if you've ever been shopping on Amazon, you've probably noticed lists with products recommended to you, such as "Customers Also Bought," "Customers Also Viewed," or "Similar Products." AI-powered recommendations are like these, except they give retailers the ability to combine an array of different factors so that they can suggest the best products tailored to an individual person. For instance, if you've been browsing for books within a specific genre, AI can locate similar readers to you and then recommend other titles based on what those similar readers have enjoyed. This level of personalization can boost sales by making relevant, timely, and context-aware product suggestions. It can help customers to discover great products they wouldn't have heard about otherwise.

AI-driven chatbots are another new technology that has already started to change the retail industry. These chatbots are able to provide instant support around the clock, which makes them a game-changer for those who need quick solutions or have trouble clearing their schedule during normal 9-5 business hours. They can handle inquiries, resolve issues, and provide information without human intervention, but there are still things they aren't capable of yet. That's why they usually work in tandem with human support representatives for more complex issues.

One example of an AI-driven chatbot is Sephora's Virtual Assistant on Facebook Messenger and Kik. Sephora's chatbot uses a series of questions to make a highly personalized product recommendation (Tiffany, 2023). However, other retailers are even using chatbots for more than just product recommendations. Very.co.uk is set up to let customers track orders with a chatbot, and Whole Foods uses chatbots to make it easier for customers to locate their stores (Kokoszka, 2018). Chatbots have the ability to learn from interactions for continuous improvement in service quality and speed. As they learn, they streamline the customer service process and allow human agents to focus more on complex customer needs.

In the present, AI is rapidly integrating with augmented reality (AR) technology, setting new standards and shifting methods of engagement within the retail sector. AR applications powered by AI can transform the way consumers interact with products before making a purchase. Virtual try-on experiences are a prime example of this integration. Warby Parker, for instance, allows customers to try on glasses at home using only a smartphone (Kristensen, 2022). Another company, Farfetch, allows customers to stand in front of a phone camera and get personalized suggestions by asking for them—the AI will use voice recognition and NLP (Natural Language Processing) to make the suggestions (Kristensen, 2022). These strategies can enhance the shopping experience, reduce the likelihood of returns, and, when relevant, reduce the need for samples to be shipped to customers so they can try them on before purchasing.

AI's integration into the retail sector will eventually further revolutionize the industry. It's likely that personalized recommendations will become more accurate, relevant, and timely, while chatbots will expand in the ways they can be used, and AR experiences will become more widespread and accessible. The future of shopping will undoubtedly be characterized by even greater personalization, efficiency, and customer-centricity, redefining the world of retail in ways that were once unimaginable.

Manufacturing

Smart Factories

In the context of industrial automation, AI has the potential to transform traditional manufacturing processes into streamlined operations that need less human oversight and intervention. AI algorithms are already capable of controlling robotic arms with meticulous accuracy, as we saw in the section on healthcare, where they performed laparoscopic surgery without human assistance. As technology evolves, it's conceivable that these technologies will widen the scope of what they can accomplish, automating industrial processes once reserved for humans like how the Industrial Revolution changed the way products are produced.

In the area of quality control, AI systems equipped with advanced sensors and vision capabilities could keep a handle on manufacturing quality. These systems would perform continuous inspections of products as they move along the conveyor belt, much like humans do now. In fact, technologies like JAKA Collaborative Robots from Industrial Automation Company already provide some features like this. According to Industrial Automation Company's website, "JAKA cobots can quickly master all applications from inspection to assembly and palletizing to welding" (Industrial Automation Co., n.d.). In 2050, I see AI getting smarter and more capable when it comes to these processes.

Like with retail, AI in manufacturing can utilize predictive analytics to forecast product demand and adjust production schedules, and therefore inventory levels, to reduce overstocks or stockouts. Additionally, AI can enhance logistics operations by determining the most efficient routing of materials and finished products, reducing delivery times and costs. AI might analyze traffic patterns and shipping statuses in real time to reroute a delivery truck, avoiding delays so that there's a timely arrival of materials or products.

It's not just about enhancing a company's bottom line. AI in predictive analytics can play a big role in enhancing worker safety in manufacturing environments. Through predictive analytics, AI systems can identify potential safety hazards before they pose a risk to workers. They do this by analyzing data from equipment sensors and operational metrics. Based on the results of these analyses, AI systems can then predict equipment failures or hazardous conditions, enabling preemptive action to mitigate risks. Furthermore, AI-driven wearables and environmental monitoring devices can track health indicators and environmental conditions, serving as more sources of data for an AI system to draw on as it determines risk.

As AI continues to permeate the manufacturing sector, a lot of growth can be expected in this area. Things are shifting toward an era where machines not only execute tasks but also oversee those tasks, inspect working environments, make predictions, and innovate solutions. In the future, this all points to an intelligent manufacturing industry that promises a safer, more efficient, and more lucrative production environment for creating better quality products. In other words, smart factories.

Financial Sector
Risk, Fraud, and Personalization

AI's integration into the financial sector is about automating tasks and transforming them into streamlined, efficient, and highly personalized processes. This transformation is evident in algorithmic trading, where AI algorithms have completely changed how some trades are executed on stock exchanges around the globe.

Algorithmic trading uses complex AI algorithms to automate stock-market trading processes in a way that makes them faster and more profitable. These algorithms are designed to analyze market conditions in real-time, interpret vast amounts of financial data, and then execute trades at optimal moments. For instance, an AI trading system might analyze trends from market data to predict a stock price increase and automatically execute a buy order. Conversely, it can detect signs of a possible price drop and sell off stocks just in time to maximize profits or minimize losses. These algorithms can enable AI systems to exploit even the smallest of market inefficiencies. Furthermore, these systems can adapt to new market conditions as they evolve, continuously learning and improving their predictive accuracies (Seth, 2023).

> ***... these systems can adapt to new market conditions as they evolve, continuously learning and improving their predictive accuracies.***

Beyond trading, AI's capability to recognize patterns and detect anomalies plays a crucial role in fraud detection and prevention. Financial institutions are already increasingly deploying AI systems that can analyze transaction data in real-time to identify behaviors that deviate from the norm, which could indicate fraudulent activities. For example, American Express uses a ML-solution called Enhanced Authorization and Chase Merchant Services uses both supervised and unsupervised learning to fight fraud (Altex Soft, 2020). By setting parameters that define typical user behavior, these AI systems can flag transactions falling outside these parameters, such as unusually large transfers or frequent transactions in a short period. Once flagged, these transactions can be halted pending further investigation, reducing the risk of financial losses due to fraud.

Credit scoring is another area within the financial sector where AI can bring substantial improvements. Traditional credit scoring methods often rely on a limited set of data points, such as loan repayment histories, and might not accurately reflect an individual's creditworthiness. AI models in credit scoring, though, can incorporate a much broader range of data, including utility bill payments, rental payment histories, and even social media activities, to paint a more complete picture of an individual's financial behavior. By doing so, these AI systems can assess credit risk with increased accuracy and inclusivity, providing credit opportunities to underserved populations who might not have a traditional credit history (Datrics, n.d.).

I assume most of your present interaction with AI in the financial sector comes from your personalized banking experiences. AI excels in analyzing customer data to understand individual preferences and behaviors, enabling banks to tailor their advice and recommendations much like retail can with their chatbots and virtual assistants. AI in finance can offer you tailored suggestions based on spending patterns and saving goals, which includes things like customized budgeting advice or suggested investment products that align with your risk tolerance and financial goals. AI-driven personal finance assistants help you manage your finances by providing timely insights and reminders, such as alerts about your upcoming bills or recommendations for ways to save money. These personalized services can enhance customer satisfaction and foster a deeper sense of loyalty and trust toward a financial institution, which is invaluable in a competitive market landscape (Google Cloud, n.d.)

The potential for AI to innovate the financial sector remains vast. From transforming how we trade and manage risk to personalizing customer interactions and expanding credit access, this ongoing transformation will lead us toward a new era of inclusivity and personalization in financial services.

Agriculture

Precision Farming and Sustainable Practices

I have never been a farmer, but it's clear agriculture holds significant promise for AI. Artificial intelligence can make farming more precise, sustainable, and productive. By working together with drones and IoT sensors, AI can transition modern agriculture into a high-tech industry focused on both efficiency and conservation, ensuring the planet's increasing food demands are met in a way that supports the health of its ecosystems.

A company like Blue River Technology is a good example. They start with crop and soil monitoring, and they use tools like Facebook's PyTorch to build machines that help farmers control weeds (Padwick, 2020). In the future, drones equipped with high-resolution cameras and IoT sensors could provide increased real-time data on crop health, soil quality, and moisture levels. The data could then be processed using sophisticated AI algorithms that identify signs of disease, pest infestation, or nutrient deficiency. Imagine drones flying over a wheat field, capturing images and data to reveal early signs of rust disease. Then, imagine they pass this data onto new drones, which use it to apply targeted treatments, limiting the spread of disease and reducing the need to apply broad-spectrum pesticides.

In fieldwork, humanoid robots are already performing labor-intensive tasks such as harvesting crops. Equipped with advanced sensors and AI, these robots can identify ripe produce and harvest it with precision. Actively deployed in crop fields today, they work alongside farmers to boost productivity, improve efficiency, and reduce operational costs.

AI's capability to predict and manage weather conditions starts to transform how farmers plan and execute their agricultural strategies. Advanced machine learning models can analyze historical weather data and current conditions to forecast weather with high accuracy. This predictive power is extremely helpful for farmers in regions prone to extreme weather events. Knowing in advance about something like a hurricane or a drought can prompt a farmer to deploy protective measures. As weather predictions become more accurate due to AI, farming processes become more strategic, efficient, and successful (Rusnak, 2024).

With the global population on the rise and natural resources dwindling, the need for efficient resource utilization is more pressing than ever. AI systems can optimize the use of water, fertilizers, and pesticides, thereby enhancing productivity while reducing environmental impact. These systems can analyze data from soil sensors and weather forecasts to determine the precise amount of water crops need, automating irrigation systems to deliver the right volume at the right time. In other words, it's about conserving water and preventing the leaching of nutrients and chemicals into waterways

Similarly, AI-driven precision spraying technology ensures fertilizers and pesticides are applied only where needed, in the exact quantities required, minimizing runoff and reducing input costs. This targeted approach can boost crop health and yield while playing a vital role in promoting sustainable agricultural practices that preserve the environment for future generations (Brears, 2024).

Yield prediction is a critical aspect of agricultural planning and market preparation. AI models have the potential to integrate various data streams, including weather data, crop health information, and market trends, to predict how much produce a particular field will yield. Through integration, farmers can better plan storage, marketing, and sales strategies well in advance, aligning supply with market demand. For example, if an AI model predicts a bumper crop of tomatoes, a farmer might arrange for additional storage facilities or negotiate contracts with processors and retailers beforehand. This foresight can prevent last-minute scrambles and loss of produce, maximizing farmers' returns on investment. Moreover, yield predictions can inform crop insurance policies and government food security strategies, making them more responsive to actual agricultural outputs.

As we continue to explore AI's applications in various industries, agriculture presents one of the most compelling cases of the power of technology. By incorporating AI to assist in monitoring soil and crop health, managing resources, monitoring the weather, and predicting yield, we can redefine our agricultural processes around the world to make them more efficient and more sustainable than ever before.

Entertainment and Content Creation

Music, Video, Art, Game, and Design

Artificial intelligence can likewise be a collaborator, shaping how music is composed, art is created, and games are developed.

It might be hard to believe, but AI software is capable of producing compositions that resonate with the depth and complexity of human-created music. These AI systems are trained on vast datasets encompassing a variety of music genres and styles, allowing them to understand and replicate complex musical structures. The process involves analyzing patterns in melody, rhythm, and harmony, enabling AI to generate new compositions that maintain stylistic consistency with existing music. Existing examples of AI's ability in music creation include the song Savages by Alltta and World is Mine by the Japanese group Supercell (jeffbullas.com, 2024). AI's music capabilities can aid musicians in exploring new creative possibilities and democratizing music creation, allowing individuals without formal training to try their hand at composition.

Visual art creation can similarly be aided by AI, which can now generate stunning artworks that are sometimes indistinguishable from those created by human artists. AI algorithms trained on diverse art styles—from classical to contemporary—can create unique pieces that reflect the intricacies and emotional expressions characteristic of human art. By inputting parameters related to style, color palette, and composition, users can guide the AI to produce artworks catering to specific aesthetics or emotional themes. One great example of this is OpenArt, which allows anyone to explore, create, and iterate with AI drawing tools (OpenArt, n.d.).

...AI can predict what kind of content will resonate with different audiences...

The same is true in the realm of video editing. AI-driven tools can automatically cut scenes, adjust sound levels, and even suggest edits based on the narrative content of the footage. One example is the software Synthesia, which allows users to create video from text (Synthesia, n.d.). This automation significantly speeds up the editing process, letting creators focus more on the creative aspects of filmmaking rather than the technical details. AI's ability to analyze video content for elements such as pacing, color grading, and audio quality keeps the final product polished and professional, aligning with the creator's vision while streamlining workflow inefficiencies.

In the gaming industry, AI can be useful in helping games to adapt to the skill level of players in real-time. It enables a challenging yet enjoyable experience for anyone playing the games. Additionally, AI is used to develop complex non-player characters (NPCs) that react intelligently to player actions, contributing to rich storytelling and interactive gameplay.

Personalized content creation extends beyond music and art into the realms of advertising and media. Here, AI excels in tailoring content to individual user preferences, enhancing engagement and satisfaction. By analyzing user data, AI can predict what kind of content will resonate with different audiences, customizing advertisements, news feeds, and even movie recommendations to align with individual tastes. This level of personalization increases the effectiveness of marketing campaigns by delivering content that is more likely to engage and convert viewers.

Similarly, by suggesting design alterations and improvements, AI tools can accelerate the creative process and enhance the aesthetic and functional quality of professional designs. Whether in graphics, architecture, or digital media, AI applications can automate routine tasks and offer solutions that optimize layout, color usage, and material selection.

Finally, the entertainment industry is witnessing the rise of AI-generated virtual performers and digital humans, which may very well transform the music and film industries. These virtual entities, powered by AI, can deliver performances that are more lifelike than ever before. One example is Miquela, a virtual Instagram model who creates music (jeffbullas.com, 2024). The integration of AI with virtual and augmented reality technologies can further enhance these experiences through the creation of immersive environments where users can interact with digital elements as if they were part of the physical world.

AI's integration into the entertainment and content creation sectors will most likely further change these industries, offering unprecedented opportunities for customization, interactivity, and creative expression. It could do a lot to push the boundaries of what is possible in art and storytelling, heralding a new era that blurs the lines between technology and creativity.

Defense

The Future of Warfare

I do not really promote the use of AI in warfare (though it can still serve well in maintaining peace), but the full picture wouldn't be complete without discussing this industry. Artificial intelligence has many military applications. It can be incorporated into autonomous weapons systems and used for advanced cyber warfare capabilities, surveillance, reconnaissance, and even for developing combat strategies.

Autonomous weapons systems powered by AI are one of the more noteworthy developments in modern military technology. These systems include AI-driven drones and robotic platforms, which are designed to perform complex tasks such as surveillance, target recognition, and even engaging targets without direct human intervention. The utilization of AI in these systems allows for deadlier, more efficient, and more accurate decision-making, which can be crucial in combat situations where every second counts (Makaryan, 2023).

In the domain of cyber warfare, AI's role is equally transformative. Cyber defense and offensive operations are increasingly driven by AI technologies that can predict, detect, and respond to cyber threats with unprecedented speed and efficiency. AI systems are trained to continuously learn from cyberattack patterns and adapt to new strategies employed by cyber adversaries. This capability strengthens the defense mechanisms of military and civilian networks while enabling the development of proactive strategies that can neutralize threats before they cause harm. These systems analyze vast amounts of data to recognize suspicious behaviors and anomalies that indicate cyber threats, then they take appropriate action (Morgan Stanley, 2023).

The field of Intelligence, Surveillance, and Reconnaissance (ISR) is being revolutionized by AI as well. AI-powered analytics can enhance situational awareness by processing and synthesizing information from diverse sources, including satellite imagery, sensor data, and intelligence reports, allowing military analysts to gain a comprehensive understanding of the battlefield so that strategic decisions can be made. Additionally, AI can improve the response time after it collects insights. Real-time decision-making support systems powered by AI further augment the capabilities of military personnel by offering actionable insights and recommendations during missions based on all this data. The result is AI and humans working together to address problems in the most intelligent way (Manzi, 2024).

> ***The result is AI and humans working together... in the most intelligent way.***

AI systems are ultimately designed to work seamlessly alongside human soldiers on the battlefield. The cognitive and analytical strengths of AI can complement the strategic and ethical judgment of soldiers and other operatives. AI-enhanced devices and vehicles can perform routine or high-risk tasks, which allows human soldiers to focus on complex decision-making and leadership tasks and can save lives. However, one big hurdle before AI can become more integrated is increasing trust in AI. In its infancy, AI may still need to learn more, grow a little further, and showcase more positive feats before people can trust it to perform as it should in important scenarios.

AI's role in defense will only grow as we progress through the 21st century. It has a ton of potential to enhance the capabilities of military forces, and it promises a future where defense strategies are more precise, efficient, and responsive. To achieve these things, we will need to ensure that the use of AI in defense adheres to ethical standards and contributes to global stability and peace.

We have just explored the impact of AI across various industries, though not all of them. From healthcare and transportation to financial services and the defense industry, AI has a myriad of applications that promise to shape a transformed world by the middle of the 21st century. Things like predictive analytics, virtual assistants, and the combination of AI with robotics and IoT devices are not going away. In chapter three, we will look at the role of AI in shaping the business sector, further demonstrating the multitude of opportunities it presents for innovation and advancement.

CHAPTER 3

AI in Business

AI in Business

AI IS REVOLUTIONIZING business sectors, acting as a catalyst for growth and transformation, from supply chain management to marketing, customer service, human resources, and even corporate decision-making processes. AI is already driving a seismic shift in how businesses operate and thrive; the pace of change is relentless.

Business Innovation

Identifying Opportunities for Growth

Consider the potential for AI to drive innovation in businesses by coming up with new ideas and strategies for growth. By analyzing trends, consumer behaviors, and market conditions, AI algorithms (with huge sets of data updated in real time) can predict shifts in markets and identify emerging niches that are ripe for exploration. They are able to analyze social media trends and online consumer feedback to help a company anticipate something like the rising demand for eco-friendly products, therefore guiding the company to pivot its product lines accordingly to capture the growing market segment.

The future hints at an era where autonomous business processes become the norm, and AI entrepreneurship shapes new industries. As its integration into business processes deepens, more sophisticated AI systems are destined to manage complex business activities with minimal human oversight. Businesses could operate on leaner models with AI handling everything from data analysis to customer interactions and even strategic decision-making. Ultimately, business ventures built around AI innovations could disrupt traditional industries with novel products and services that offer unprecedented efficiency and customization.

As AI redefines the landscape of business innovation, understanding its potential is crucial for anyone involved in the business world, whether you're an entrepreneur, a corporate executive, or a policymaker. The insights gained from AI can drive growth and sharpen competitive edges, making businesses more adaptable and resilient.

Let's have a deeper look at the specific applications of AI in business, exploring how AI can reshape industries and redefine the future of work. There's no way we can cover every application, but these are some that stand out to me. The list grows with every day.

AI Startups

Innovation and Challenges

AI startups are at the forefront of developing cutting-edge AI technologies, and they also play a crucial role in applying these innovations across various industries, from healthcare to automotive. There are a diverse variety of AI startups out there, with ventures focusing on everything from fundamental AI research and development of intelligent machine learning algorithms to specific applications such as autonomous driving, predictive analytics, and personalized medicine.

Venture capital plays a pivotal role in this ecosystem, fueling the growth of AI startups in the early stages. Recent trends indicate a significant surge in investments directed toward AI startups that integrate AI to enhance cybersecurity, improve financial services, or revolutionize supply chain logistics.

Success stories of AI startups abound (though beware, many more have failed). Take, for instance, Medial Earlysign, a 10-year-old company that has developed an AI-powered diagnostic tool leveraging big data to detect diseases from medical imaging faster and with greater accuracy than traditional methods. This innovation increases the chances of successful treatment by catching diseases in their early stages. Medial Earlysign is currently targeting lower GI disorders, prediabetic progression to diabetes, chronic kidney disease, and coronary artery disease. Their latest products, which use a tool called Algomarkers, have already outperformed methods that came before in detecting prediabetic patients with the highest risk of progressing to diabetes and type 2 diabetics with the highest risk of progressing to chronic kidney disease (Nanalyze, 2019).

The journey of AI startups is fraught with challenges. Securing adequate funding is often the first major hurdle, as investors may be hesitant to back ventures involving high risks or long development periods typical of AI projects. Navigating the regulatory environment poses another challenge, especially for startups in sectors like healthcare or financial services, where compliance with stringent regulations is mandatory. Additionally, the scarcity of talent in the AI field can impede a startup's ability to innovate and scale. The demand for skilled AI professionals far exceeds the supply, making it difficult for startups to attract and retain the talent needed to drive their development forward.

The landscape for AI startups is both promising and competitive. The increasing integration of AI with other emerging technologies, such as blockchain and the Internet of Things (IoT), should spawn a new wave of startups that could disrupt traditional business models and create new market niches. Companies like WINT, which uses AI in combination with water meters to create smart meters that can detect anomalies in water flow (Patel, 2024), and Bext360, which uses AI in combination with the blockchain to boost supply chain transparency and efficiency across a range of industries (Daley & Whitfield, 2023), are already making big strides in these integrations.

In the future, new emerging forces might reshape the AI startup ecosystem. The increasing scrutiny of AI technologies by governments and regulatory bodies could result in new regulations that might constrain the operations of AI startups. In addition, as AI becomes more mainstream, larger tech companies may enter the market, intensifying competition and possibly leading to consolidation in the industry.

Supply Chain Management
A Revolution in Logistics

In the supply chain, AI is already enhancing existing processes and redefining them—creating smarter, more resilient, and highly efficient supply chains. Consider the optimization of logistics, where AI can transform how products are stored, moved, and delivered. By using AI to optimize routes and manage inventory levels, businesses can achieve a remarkable reduction in operational costs while majorly improving delivery times. AI algorithms can analyze vast datasets involving traffic patterns, weather conditions, vehicle conditions, and driver performance to determine the most efficient routes. This minimizes fuel consumption, so that goods are delivered in the most timely and cost-effective manner possible. Moreover, AI-driven systems can dynamically adjust routes in real-time, responding to unexpected road closures or logistic challenges, thus maintaining the fluidity of supply chains.

AI is an indispensable tool in demand forecasting. Traditional forecasting methods often rely on historical sales data, but AI introduces a more nuanced approach by incorporating a wider array of variables, including market trends, consumer behavior, economic indicators, and even social media sentiment. This holistic approach allows for predictions that are more accurate and more granular, enabling businesses to anticipate demand fluctuations with a high degree of precision. An AI system might predict an increased demand for certain products based on upcoming weather conditions, allowing retailers to adjust their inventory in advance. This predictive capability is invaluable in industries where products have a limited shelf life, such as food and pharmaceuticals; it reduces waste and ensures the availability of stock, thereby enhancing customer satisfaction.

Supplier relationship management is another area where AI makes a big impact. In today's global economy, managing supplier relationships effectively is crucial for maintaining the quality and reliability of supply chains. AI tools streamline the process by assessing supplier performance data to identify risks and opportunities. These tools can evaluate factors such as delivery times, quality incidents, and compliance with contractual agreements to provide a comprehensive assessment of supplier reliability. Afterward, AI can help in selecting new suppliers by analyzing global data to identify partners who meet specific criteria related to sustainability, cost, and delivery capabilities. This maintains a high standard of procurement and fosters a competitive spirit among suppliers to maintain their performance levels.

Resilience building in supply chains has never been more critical, especially in light of recent global challenges such as the COVID-19 pandemic. AI contributes to enhancing the resilience of supply chains by enabling companies to predict, respond to, and recover from these disruptions. Advanced AI models can simulate various disruption scenarios and their impacts on the supply chain, allowing companies to develop contingency plans and proactive strategies. For example, AI can identify alternative suppliers or logistics options when a primary supplier is compromised, ensuring that the supply chain continues to function with minimal impact. Additionally, AI-driven analytics can monitor global events in real time, providing early warnings of possible disruptions such as natural disasters, strikes, or political unrest, allowing companies to react swiftly and mitigate risks (Javaid, 2024).

This predictive capability... reduces waste and ensures the availability of stock, thereby enhancing customer satisfaction.

The integration of AI technologies into supply chain management is likely to intensify, leading to additional innovations and greater efficiency. The future shows the potential for even more sophisticated AI applications, from autonomous vehicles conducting deliveries to blockchain-enhanced transparency and security in logistics. These technologies hold the key to transforming supply chains into seamless, automated, and highly resilient systems.

By optimizing logistics, enhancing demand forecasting, improving supplier relationships, and building resilience, AI streamlines supply chain operations and equips businesses with the tools to navigate the complexities of global trade. The potential for innovation and transformation in this sector is immense.

Marketing
Personalization and Consumer Behavior

Marketing is no longer just about reaching the widest audience possible but about reaching the right audience with precision and personalization. Artificial Intelligence has become a linchpin in this shift, leveraging vast amounts of consumer behavior data to tailor marketing efforts to individual preferences and behaviors. By analyzing data points from online browsing habits to purchase history, AI enables marketers to craft personalized experiences that resonate deeply with consumers. This level of personalization is evident in targeted advertising, where AI algorithms can display ads that are both relevant to the consumer's recent searches and aligned with predicted needs based on their digital footprint. For example, someone browsing for hiking gear might receive targeted ads for trail recommendations or upcoming outdoor events, enhancing the relevance and engagement of the ads.

AI is capable of improving customer engagement with sophisticated tools and techniques that analyze and predict consumer behavior. Sentiment analysis, a key tool in AI's arsenal, examines the emotions behind consumer interactions on social media and reviews, providing brands with insights into customer perceptions and areas for improvement. This analysis helps companies adjust their strategies in real-time, responding to customer feedback with changes that enhance satisfaction and loyalty. Predictive analytics further refine customer engagement by forecasting future buying behaviors based on past actions. As a result, companies can anticipate customer needs and craft marketing messages that are likely to elicit positive responses, thereby aiding in customer retention.

The impact of AI in marketing is not just theoretical but is evidenced by numerous success stories where brands have harnessed AI to achieve remarkable results. Tastewise, a consumer data platform powered by

Generative AI offers tools that provide marketing insights and give companies a way to automate and execute marketing initiatives. Its solutions have already enhanced the marketing efforts of a wide range of food and beverage brands, including Nestle, Kraft&Heinz, Campbell's, and Pepsico (Tastewise, n.d.).

The future of marketing with AI appears even more promising, with innovations geared toward hyper-personalized campaigns and automated customer interactions. Hyper-personalization goes beyond traditional segmentation, using AI to create a unique marketing mix for every individual, including tailored messages, recommended products, and personalized shopping experiences. These AI-driven strategies can ensure that marketing efforts are more effective and efficient, providing substantial returns on investment.

Ultimately, by leveraging AI to understand and predict consumer behavior, personalize marketing efforts, and enhance customer engagement, companies can achieve better marketing outcomes and build stronger, more meaningful relationships with their consumers. The era of AI-driven marketing is already here, and it offers a new way for brands to interact with their markets, elevating personalization in incredible ways.

Customer Service Automation

Enhancing Efficiency and Satisfaction

In the realm of business technology, artificial intelligence has begun to revolutionize customer service, transforming it from a traditional reactive model to a dynamic, interactive, and satisfying experience. The deployment of AI through chatbots and virtual assistants epitomizes this transformation. These AI-driven tools are now sophisticated interfaces that manage customer inquiries with an efficiency and consistency human agents often struggle to match. They operate around the clock, provide immediate responses to customer queries, and attend to each and every customer interaction, regardless of the hour. It is also a big boon for companies that serve customers who struggle to be available during normal business hours or have problems requiring a prompt response.

Personalization stands at the core of modern customer service, and AI is pivotal in delivering personalized experience. By analyzing data from past interactions and customer behaviors, AI tools can tailor their responses to meet the specific needs and preferences of individual customers. If a customer has a history of purchasing eco-friendly products, the AI tool can prioritize such a preference in its recommendations and queries. It then can enhance customer satisfaction and loyalty.

Integrating AI tools with human customer service teams presents a unique set of challenges and opportunities. Best practice for this integration focuses on leveraging AI to handle routine queries and data-heavy tasks, allowing human agents to focus on areas where human empathy and complex problem-solving are required. A collaborative approach ensures AI handles only the things it does well while human employees fill in the gaps. Training staff to work alongside AI tools and understand their capabilities and limitations is crucial. Equally important is the establishment of clear protocols for handing off issues from AI systems to human agents when the interaction surpasses the AI's ability to resolve effectively.

Human Resources

Recruitment and Management

Through AI, the recruitment process can transition from a labor-intensive task to a streamlined operation characterized by efficiency and precision. AI excels in parsing through data, such as thousands of jobs applications to identify the most suitable candidates. It achieves its results through sophisticated algorithms capable of analyzing a vast array of data points—from past job experiences and educational backgrounds to skill sets and even subtle indicators like cultural fit. These systems are not merely filtering based on keywords but are intelligently matching candidates with job profiles using predictive analytics. Ultimately, this reduces the likelihood of turnover and enhances the quality of hires.

In the areas of employee management and development, AI systems can be utilized to tailor training programs to individual employees. By analyzing performance data and learning patterns, AI can identify skill gaps and learning preferences, allowing organizations to offer personalized training that is both effective and engaging. Personalization boosts employee skills and enhances job satisfaction and loyalty. AI-driven career development tools can further support employees by suggesting pathways and opportunities within the organization that align with their skills and aspirations. This helps in retaining top talent, as it shows a clear trajectory for career advancement, making employees feel valued and invested in their roles.

One great example of an AI tool that can help with employee training is FutureFit AI. FutureFit works alongside multiple different Fortune 500 companies to help employees navigate their careers (FutureFit AI, n.d.).

Ultimately, this reduces the likelihood of turnover and enhances the quality of hires.

However, the integration of AI in recruitment and management can bring difficulties regarding bias and fairness. Despite the efficiency of AI, there's an inherent risk that these systems might perpetuate existing biases present in the data they learn from. If historical hiring data shows an unintended bias toward a certain demographic, AI systems are likely to inadvertently continue this trend unless explicitly designed to counteract such biases. Such was the case with one of Amazon's AI recruitment tools back in 2015, which was largely trained on hiring data from male employees, who dominate the tech industry. As such, the system developed an unintended bias towards men over women (Dastin, 2018). In the long run, addressing the issue involves a combination of rigorous algorithm training, regular audits, and the integration of fairness modules to design AI recruitment tools that treat all candidates equitably.

AI's ability to affect recruitment and management is groundbreaking. By streamlining recruitment processes, personalizing employee development, practicing fairness in hiring, and providing actionable insights into workforce dynamics, AI tools can become indispensable in modern HR practices. These advancements will play a crucial role in shaping a work environment that is dynamic, inclusive, and aligned with the strategic goals of an organization. They create a workplace culture that is adaptive, innovative, and forward-thinking.

Corporate Decision-Making

Enhancing Analytical Capabilities

By harnessing the power of AI, businesses can gain access to data-driven insights and predictive analytics that illuminate current conditions and forecast future trends. Having insights like these enables leaders to make decisions that are proactive rather than reactive.

In essence, AI enhances decision-making by integrating diverse data sources—market trends, consumer behavior, financial indicators, and competitive analysis—into a coherent framework that provides a holistic view of the business environment. This allows AI systems to identify patterns and correlations that might be invisible to even the most skilled analysts. For example, AI can pinpoint emerging market opportunities by analyzing shifts in consumer preferences and demographic trends, or it can highlight risks from economic downturns or changes in regulatory landscapes. These insights empower executives to make informed decisions that align with both short-term realities and long-term visions.

AI systems can use predictive analytics to forecast future trends, enabling businesses to anticipate market movements, consumer behaviors, and potential disruptions. All the predictive power can turn uncertainty into an array of calculated risks and informed choices.

In the future, AI's role in enhancing corporate decision-making will grow, with newer models offering even greater accuracy, speed, and contextual awareness. These advancements will further empower executives to navigate the complexities of the global market with confidence, armed with insights that both respond to current conditions and anticipate future challenges and opportunities. In this way, AI supports businesses in transforming problems into opportunities for growth and innovation.

The integration of AI into business strategies is not just beneficial but essential for any organization aiming to succeed in the modern economy. Embracing AI in decision-making processes offers a significant strategic advantage, and the gap in results between those who do this and those who don't will only become more pronounced as time goes on.

CHAPTER 4

AI in Technology and Innovation

AI in Technology and Innovation

FROM GENERATIVE AI tools like GPT-4, chatbots, and virtual assistants to AI and the blockchain, integration with IoT, renewable energy collaborations, and even applications in genetic engineering, AI has the potential to reshape and redefine a variety of different industries. This chapter is about both understanding the mechanics of the collaborations between AI and these technologies and envisioning the growth needed to reshape societal interactions. It's about dispelling traditional barriers and forging new pathways in a secure, smart, and sustainable world.

Generative AI in Communication

Generative AI, a frontier within artificial intelligence, involves technologies capable of producing content, whether textual, auditory, or visual, that is indistinguishable from content generated by humans. At the heart of this revolution is GPT-4, a model that exemplifies the culmination of advancements in natural language processing. GPT-4 and its predecessors are trained on diverse internet text, but the sophistication with which GPT-4 understands and generates human-like text is unprecedented. It interacts contextually, adjusts style and tone adeptly, and generates complex written content that resonates with human emotions and intellectual nuances.

The applications of such generative AI in communication are profound and multifaceted. In customer service, AI-driven chatbots equipped with capabilities derived from models like GPT-4 are transforming interactions. They are no longer mere responders to queries but are now capable of leading conversations, providing personalized customer experiences, and even managing crises, all in real time. The era of waiting in queues for human agents is being eclipsed by instant, efficient, and empathetic communication managed by AI.

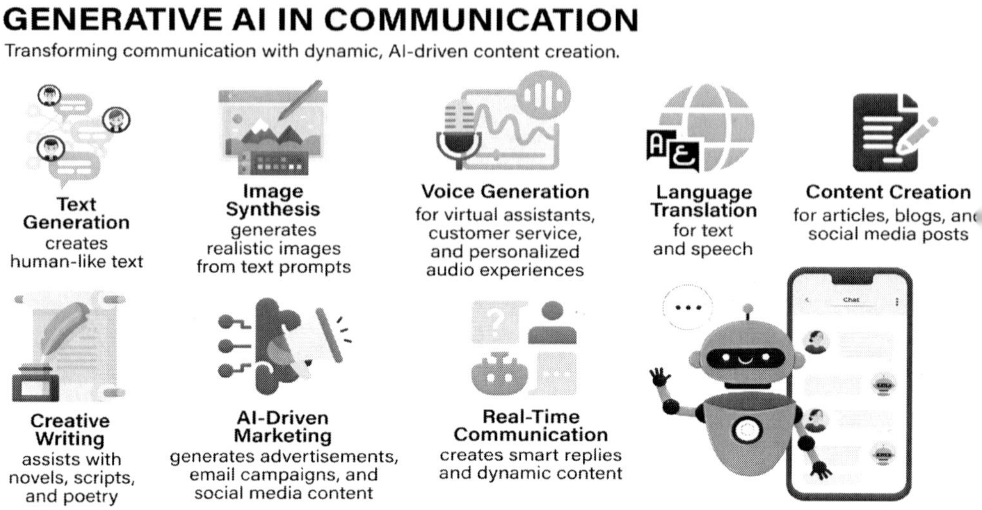

Generative AI is breaking down long-standing language barriers, such as in the case of Jugalbandi, a multilingual chatbot, which was used to help speakers of Telugu in India get information on government aid and help women and children to deal with harassment and domestic violence (Reynolds, 2024).

Imagine a small business owner in New Zealand (like me) reaching customers in a niche market in rural Japan; generative AI can make this possible through real-time, accurate translation that captures the language and cultural nuances essential to effective communication. This democratization of language services expands markets, brings the world closer together, and fosters understanding and cooperation across borders.

Content creation, too, has been revolutionized by generative AI. From generating news articles and writing scripts to composing poetry and novels, AI is capable of tapping into creative realms that were once considered solely human domains. Many of these revolutionary technologies are based on OpenAI's ChatGPT, which we will discuss more in a moment. For example, Jasper, a marketing-focused technology based on GPT-3, is capable of all sorts of content from blogs to social media posts, advertisements, and even sales emails (Davenport & Mittal, 2022). Ultimately, this technology is not about replacing human creativity but about augmenting it, providing tools that can inspire and assist human creators to push the boundaries of what's possible in art, literature, and beyond.

Case Study: ChatGPT

A practical example of generative AI in action is ChatGPT, developed by OpenAI. It's a conversational model trained on a mixture of licensed data, data created by human trainers, and publicly available data. Designed to follow instructions in a given prompt and provide detailed responses, ChatGPT has rapidly gained popularity, boasting over 200 million monthly active users globally as of 2024 (SimilarWeb, 2024). What sets ChatGPT apart is its ability to understand the context of a conversation, making it an invaluable tool for applications ranging from educational tutoring to interactive storytelling. In customer service, ChatGPT provides a model for the future, showcasing how businesses can automate responses without losing the human touch that is so crucial to customer satisfaction.

ChatGPT's adaptability allows it to be fine-tuned for specific industries. In healthcare, it can assist medical professionals by quickly accessing and summarizing patient records, providing reminders for medication schedules, and offering preliminary diagnoses based on symptom descriptions. In the financial sector, ChatGPT can enhance client interactions by answering queries about account balances, transaction histories, and investment advice, all while maintaining stringent privacy standards. Another notable application is in content creation, where it aids writers and marketers by generating creative ideas, drafting articles, and even editing text for clarity and coherence.

Future Trends

In the future, the trajectory of generative AI in communication points to even more integrated and intuitive applications. Emerging trends suggest the development of AI models that can better understand and mimic human emotions, leading to more sophisticated interactions where AI can act as a mediator in negotiations, a collaborative partner in creative projects, or a personalized companion in daily activities. Furthermore, the integration of AI in virtual reality settings may provide a future where AI-generated content is interactively experienced, creating immersive environments that could be used for everything from education to therapeutic sessions.

All in all, the implications of these technologies extend beyond practical applications; they invite us to reimagine the very nature of how we connect, understand, and interact with one another. In the new era, where communication is unbounded by language barriers or creative limitations, the possibilities are as vast as our collective imagination allows them to be.

AI and Blockchain

Enhancing Security and Transparency

AI's ability to analyze vast datasets can be combined with the blockchain's inherent characteristics of decentralization and tamper-resistant records. One of the primary strengths of blockchain technology is its capacity to create immutable and transparent records of transactions. However, as blockchain systems become increasingly complex and widely adopted, they face some security risks. It's on this front that AI can step in to provide additional functionality and support. By employing machine learning algorithms, AI can continuously monitor blockchain networks for unusual patterns or breaches. For instance, in cases of security threats like double-spending attempts or the emergence of suspicious blocks, AI systems can alert administrators and even automatically initiate protocols to thwart these threats.

Transparency, a cornerstone of blockchain technology, can be further enhanced by AI's proficiency in processing and analyzing transactions at exceptional speeds. AI can facilitate real-time auditing of transactions and provide all parties with immediate insights into the state of their operations and financials. This level of transparency is revolutionary in industries like supply chain management, as it enables stakeholders to trace the provenance and status of goods instantaneously, keeping things compliant and authentic.

Smart contracts—self-executing contracts with the terms of the agreement directly written into code—are another area where AI has contributed meaningfully to blockchain technology. Traditional contracts often require human oversight for execution and conflict resolution, which can lead to delays and increased costs. AI, though, can predict outcomes based on historical data and external variables, making it possible to automate adjustments and even resolve disputes. By integrating AI with smart contracts, the execution becomes faster and more reliable, as AI's predictive capabilities can identify issues before they arise, ensuring that the terms are adjusted or enforced accordingly.

CertiK, a blockchain company based in New York, is a great example of a company using AI to enhance blockchain products. CertiK's tools utilize AI to identify security risks, monitor data insights, and help secure blockchain, smart contracts, and Web3 applications (Daley & Whitfield, 2023).

Future Applications

The intersection of AI and blockchain holds promise for fully autonomous decentralized organizations (DAOs). These digital entities operate without the need for traditional management structures and governed entirely by smart contracts on a blockchain. AI could drive these organizations, making decisions based on predefined criteria and real-time data analysis, redefining corporate structures while reducing overhead costs and increasing efficiency. However, the realization of such sophisticated DAOs will first require advancements in both AI decision-making capabilities and blockchain scalability—challenges that researchers and developers are actively working to overcome.

All in all, the implications of these technologies extend beyond practical applications; they invite us to reimagine the very nature of how we connect.

With AI evolving in its ability to handle complex decision-making and problem-solving, its integration with blockchain will likely lead to innovations in various sectors. In finance, AI-enhanced blockchain systems could automate most banking services, providing customers with faster, more secure financial transactions. In governance, blockchain can ensure the integrity of electoral processes, with AI analyzing voting patterns to detect fraud and keep processes fair.

These are only two of the many potential use cases for these technologies together.

Integration of AI with IoT

Smart Everything

The fusion of artificial intelligence and the Internet of Things (IoT) can transform how our environments—from homes and offices to entire cities—respond to and anticipate our needs. AI enhances IoT devices with sophisticated decision-making capabilities. Think of something like Amazon Alexa, which can communicate with you using Natural Language Processing to do tasks or retrieve information. Imagine its capabilities expanded to other devices that you interact with on a day-to-day basis. This could be your living room, office, or even your city performing tasks and understanding and adapting to your preferences and habits, optimizing your environment for comfort, efficiency, and sustainability.

This transformation is grounded in the capability of AI to infuse IoT devices with smart decision-making. Through the continuous collection and analysis of data from sensors and devices, AI algorithms can learn patterns and make decisions that optimize the functionality of these devices. For instance, smart thermostats in homes and offices can adjust heating and cooling systems based on your daily routines and weather predictions, optimizing comfort while minimizing energy use. Similarly, in smart cities, IoT sensors can monitor traffic flow and coordinate with AI systems to adjust traffic signals in real time, easing congestion and reducing commute times.

The optimization of resource use is another benefit of integrating AI with IoT. By analyzing data collected from various sensors, AI algorithms can manage and optimize energy consumption in real time. In the area of smart grids, for example, AI can predict peak load times and adjust energy distribution accordingly to prevent overloads and enhance efficiency. This extends to smart homes, where AI can control lighting, heating, and other home systems based on occupancy and personal preference, reducing unnecessary energy consumption and lowering utility costs. Moreover, AI-driven IoT applications in waste management can monitor waste levels and optimize collection routes and schedules, improving efficiency and prioritizing environmental sustainability.

Enhancing Human-Machine Interactions

Thanks to the integration of AI with IoT, the interface between humans and machines is undergoing a big transformation. This synergy makes interactions with technology more intuitive, seamless, and aligned with natural human behaviors. AI algorithms process data from IoT devices to learn user preferences and behaviors, allowing these devices to respond to unspoken commands or anticipate needs. For example, voice-controlled assistants who understand context and nuance can manage everything from your schedule to your smart home devices, making technology an unobtrusive and supportive part of your life.

In healthcare, AI-enhanced IoT devices can provide personalized care that adapts to patient needs. Wearable devices, such as the Apple Watch with the Apple Health app, monitor health metrics such as heart rate and sleep patterns and can alert users and healthcare providers to potential health issues before they become critical. The result is improved individual health management and the alleviation of burdens on healthcare systems by reducing the number of emergency visits and hospitalizations.

Meanwhile, in elder care, AI-integrated IoT devices can help seniors live independently longer, with systems designed to monitor safety, manage medication, and maintain communication with caregivers, enhancing both personal security and quality of life. One great example is the Kanega Watch, developed by Jean Anne Booth and her company, Unaliwear. This watch was built with Booth's 80-year-old mother in mind, and it has a built-in accelerometer to detect falls, discreet medication reminders, and waterproof casing, among other features (Jamrozik, 2019).

IoT in the Future

We can envision a future where AI-integrated IoT devices create environments that are smart, empathetic, and adaptive to the needs of their users. In this future, smart homes could evolve into dynamic ecosystems capable of supporting our daily lives, learning our preferences, and adapting to changes in our routines to optimize our living environments. Here, everything from lighting, temperature, and security to how appliances operate is managed seamlessly by AI, providing a living experience that is truly personalized and responsive.

In urban environments, the integration of AI with IoT has the potential to transform cities into interconnected hubs that enhance the quality of life for all residents. Smart city solutions may one day manage everything from traffic and waste management to energy use and environmental monitoring, creating efficient, sustainable, and livable urban spaces. In these cities, AI-driven IoT systems provide real-time data to help city managers and policymakers make informed decisions that promote safety, efficiency, and sustainability.

Renewable Energy

Optimizing Green Technologies

Artificial intelligence has many applications with renewable energy technologies. Together, these two advancements are revolutionizing how we harvest, manage, and utilize clean energy. AI's capability to forecast and optimize energy production from renewable sources such as solar and wind is instrumental in elevating the efficiency and reliability of green technologies. By accurately predicting energy output, AI enables utility providers to manage energy distribution more effectively, ensuring stability across the grid while minimizing reliance on non-renewable energy sources.

This predictive ability is another example of predictive analytics in action. Renewable energy sources are often variable in nature, affected by things like the strength of the wind over a given period of time or the availability of direct sunlight. Knowing when the energy will be most available allows smart systems to make choices that best utilize and conserve energy. For solar energy systems, for instance, AI leverages data from weather satellites and ground sensors to predict sunlight patterns, optimizing energy capture and reducing wastage. Similarly, in wind energy systems, AI algorithms analyze meteorological data to predict wind patterns, enabling turbines to adjust their angles for maximum energy generation. These changes ensure a more stable and reliable energy supply, facilitating a smoother integration of green technologies into the existing energy infrastructure.

Another critical application of AI in renewable energy is in the operation of solar power installations. By dynamically adjusting the angle of solar panels throughout the day to align perfectly with the sun's position, AI can maximize the capture of solar energy. This is achieved through sophisticated tracking systems that use AI to calculate the optimal orientation of panels at different times of the day, considering factors such as the angle of the sun and seasonal variations in how sunlight reaches the panels (Horwood, 2023).

Future Innovations

One of the most anticipated developments in the area of renewable energy is the advent of fully autonomous energy grids controlled entirely by AI. These smart grids would manage the distribution and storage of energy, and autonomously resolve issues such as repairs and maintenance, effectively eliminating downtime and reducing operational costs. The AI in these systems would continuously learn and adapt, improving efficiency and responsiveness over time, thus meeting the energy needs of consumers with unprecedented precision and reliability.

In addition, the integration of AI with advanced battery technology could revolutionize energy storage solutions. AI could optimize charging and discharging cycles based on predictive analytics, extending battery life and enhancing the efficiency of energy storage. The result would be a balanced grid during periods of high demand or low supply.

AI-driven renewable energy systems that generate and distribute power while actively participating in environmental conservation are a future possibility. AI could optimize the operation of renewable installations to minimize their ecological footprint, for example, by reducing bird collisions with wind turbines through predictive behavior modeling and automated turbine shutdowns.

Ultimately, a sustainable energy future is intricately linked with advancements in technology. AI is likely to pave the way for innovative solutions that could redefine how energy is produced, managed, and consumed globally. The journey toward a renewable, AI-driven energy economy promises environmental, economic, and social benefits that have the potential to transform many facets of our society.

AI in Weather Forecasting and Climate Change Mitigation

As our planet faces unprecedented climate challenges, artificial intelligence can be a leading force in climate change mitigation. AI's predictive modeling capabilities are already increasingly being harnessed to anticipate and manage the impacts of climate change, and this offers hope for growth and more effective strategies to protect our environment in the future. By employing sophisticated algorithms, AI systems can analyze vast amounts of climate data, from temperature patterns to atmospheric conditions, to predict climate trends and extreme weather events with remarkable accuracy. Predictive power enables governments and organizations to prepare more effectively for disasters, minimizing human and economic losses. Furthermore, AI's ability to monitor environmental changes in real time supports proactive responses to ecological disturbances, such as deforestation and ice melt, ensuring that interventions are timely and informed by the latest data.

Tomorrow.io's Gale software is one tool that is currently revolutionizing weather forecasting. Gale is designed to use ML-powered algorithms to analyze weather patterns and make predictions about future weather. Tomorrow.io describes the software as the world's first Weather AI, and it doesn't just predict the weather—it also can generate customized weather reports for professionals in varying industries. This means people in aviation, for example, can get weather reports tailored to their unique circumstances, while others in trucking or construction can get different kinds of weather reports (Favela, 2024).

AI systems like Gale integrate multiple data sources and continuously learn from new information, which enhances the accuracy and reliability of weather predictions, allows for better-prepared communities, and leads to more effective deployment of resources. AI's advanced and customized forecasting models can be instrumental in agriculture, where knowing the likelihood of weather conditions such as frost or drought can help farmers make informed decisions about planting and irrigation, thus minimizing crop losses and boosting productivity.

Ultimately, a sustainable energy future is intricately linked with advancements in technology.

Tackling carbon emissions, one of the primary drivers of climate change, is another area where AI is making significant strides. Through the development and implementation of strategies that leverage AI, industries can monitor and reduce their carbon footprint more effectively. AI systems can analyze energy consumption patterns and identify inefficiencies, suggesting changes that reduce energy use and emissions. In manufacturing, for instance, AI can orchestrate the most energy-efficient sequences of machine operations, cutting energy costs and emissions. Meanwhile, by controlling the variables involved in carbon capture processes, AI can improve the efficiency and scalability of carbon capture technologies, making it feasible to remove greater amounts of carbon dioxide from the atmosphere. Together, these two applications can get us closer to being a carbon-neutral society.

By enhancing predictive models, improving the efficiency of renewable energy sources, and developing effective carbon reduction strategies, AI serves as a powerful tool in our global efforts to safeguard the planet. The potential for AI to make environmental impacts continues to grow, and as time goes on it will contribute to both mitigating climate change and managing natural resources more efficiently.

AI in Space Exploration

One of the primary applications of AI in space exploration is the automation of spacecraft. Traditional space missions have been heavily reliant on human control, requiring continuous monitoring and intervention from mission control centers on Earth. However, AI introduces a new level of autonomy by enabling spacecraft to perform critical tasks independently. For instance, SpaceX's Falcon 9 craft uses an AI autopilot to dock with the International Space Station (Marr, 2023). AI systems on spacecraft can now navigate through space with minimal human input, using advanced algorithms to process data from onboard sensors and adjust trajectories in real-time. This capability reduces the reaction time to unforeseen circumstances and lessens the burden on human operators, allowing them to focus on more strategic aspects of the mission.

AI can play a crucial role in the health and functionality of spacecraft systems through system monitoring. With continuous data analysis, AI can predict system failures before they occur, enabling proactive maintenance and adjustments. This is vital in space missions, where repair options are limited and system failures can have catastrophic consequences. For example, AI algorithms can monitor the health of a spacecraft's propulsion system, analyzing patterns in engine performance to identify signs of wear or malfunction. By addressing these issues early, AI helps prevent failures that could jeopardize the mission, enhancing the safety and reliability of space exploration.

Similarly, AI can help spacecraft to avoid collisions with satellites, asteroids, or other objects in space. SpaceX uses AI algorithms toward these ends, and the UK Space Agency has developed its own autonomous systems to avoid debris in space (Marr, 2023).

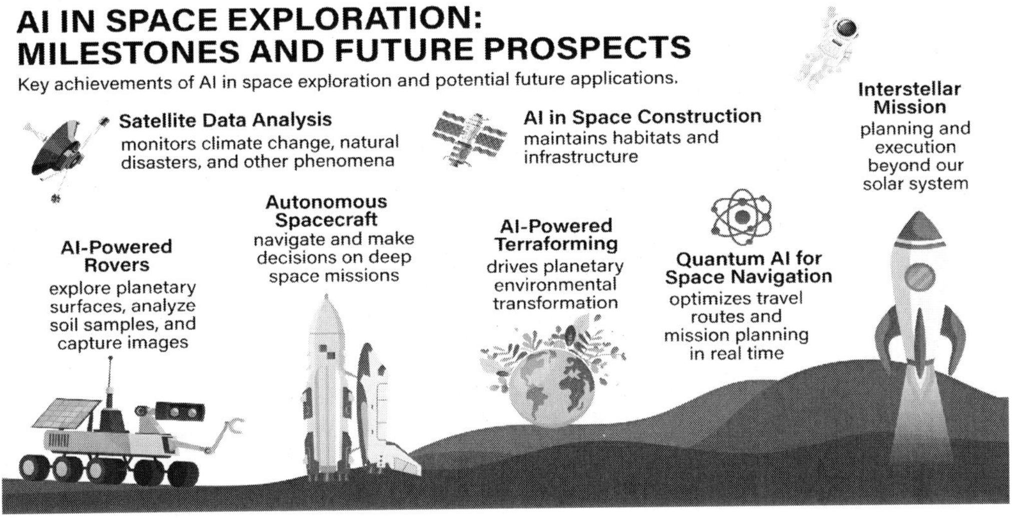

Robotics in Space Construction

In the area of space infrastructure development, AI-powered robots could play a big role. The construction of structures in space, such as satellites and space stations, has traditionally been a costly and risky endeavor, often involving complex spacewalks performed by astronauts. One day, AI-driven robots could perform construction tasks autonomously or with minimal human oversight, saving humans the risk and reducing costs. Efforts to make robots for these purposes are already underway, with projects like the Robonaut project by NASA, which aims to develop humanoid robots capable of working alongside astronauts or taking over hazardous tasks during space missions (NASA, n.d.).

AI in Space Exploration

AI-driven autonomous spacecraft can revolutionize our ability to explore the cosmos. Equipped with sophisticated AI systems, they're designed to conduct missions based on pre-programmed objectives and real-time data. Our Mars rovers are great examples of autonomous robots that assist in space exploration. As they traverse the surface of Mars, they use AI systems to monitor for and avoid craters without the need for humans to step in and direct them. In the future, as we set out on more deep-space missions where communication delays with Earth can make real-time human oversight impractical, AI will most likely become even more useful (Marr, 2023).

Mars rovers equipped with AI systems can use image recognition tools or computer vision technology to take pictures of the surface and analyze the composition of different rocks and surface features and materials (Marr, 2023). This image recognition capability aids in building a map of Mars' surface so that we can better understand exactly what it looks like without needing to handhold robots through each step of the process.

Furthermore, astronomers can use AI to aid in mapping the universe by identifying patterns in star formations or other features. NASA's James Webb telescope, launched in 2021, is already using AI in this way. Data and images from the telescope have been sent back to us here on Earth and fed into UC Santa Cruz's Morpheus system to analyze (Wodecki, 2022).

As time goes on, AI is expected to become a cornerstone of future space activities, driving innovations that expand our capabilities and deepen our understanding of the universe. From autonomous robots and spacecraft to AI technologies that can aid in better mapping the universe, the integration of AI in space exploration makes missions more efficient, safer, and enables more ambitious explorations that were once beyond our reach.

Genetic Engineering

By sifting through and analyzing vast arrays of genetic information, AI offers the promise of identifying patterns and mutations that are often subtle and elusive to human researchers. These identifications could be crucial in unraveling the genetic factors behind complex diseases such as cancer, Alzheimer's, and other genetic disorders. Understanding these factors can lead to groundbreaking preventive measures and targeted treatment plans precisely tailored to individual genetic profiles.

Personalized medicine is one promising advancement in the future. As discussed in chapter two, it can be particularly effective in treating chronic diseases like cancer and diabetes. Analyzing a patient's genetic information can give insights that lead to a customized medical treatment addressing not just the symptoms being experienced in the present, but also the root genetic causes of a condition faced now even if there are not symptoms. By 2050, it's conceivable that diseases once considered untreatable may enter the realm of management and cure thanks to personalized therapies guided by AI-driven insights..

However, the rapid advancements in AI and genetic engineering bring forth major ethical implications that must be carefully considered. The power to alter genetic material can lead to questions about the extent to which this technology should be used. Concerns about gene editing, particularly the potential to create 'designer babies' by selecting or altering genes related to traits such as intelligence, appearance, and disease predisposition, raise ethical dilemmas about inequality and the fundamental nature of human experience. These worries will need to be listened to and addressed as the technology advances. The possibility for genetic inequality, where only those who can afford gene-editing technologies have access to these advancements, poses a risk to social equity and justice. In an ideal world, all of these moral issues will be addressed before the technology develops.

Regulatory Challenges

The current frameworks governing genetic research were not designed to keep pace with the rapid advancements brought about by AI. As such, there is a critical need for updated regulations that adequately address the new realities of genetic engineering. These frameworks must monitor genetic editing technologies to ensure they are used responsibly, with clear guidelines to prevent misuse and protect against unintended consequences. For instance, regulations could specify the conditions and purposes for which genetic editing is permitted, such as for therapeutic but not enhancement purposes. Additionally, there is a need for international cooperation in developing these regulations. Genetic engineering technology raises questions that transcend national boundaries, and therefore, a global approach to governance is essential to manage the ethical, social, and legal implications associated with these technologies.

The integration of AI in genetic engineering, therefore, offers many possibilities and possible issues. Personalized medicine along with the promise for early disease detection and treatment look particularly promising, but these things will need to be balanced with new ethical and regulatory frameworks to account for the societal risks. The decisions we make today will likely shape the future of medicine and how gene editing affects the people and society of the future.

CHAPTER 5

AI in Regulation and Governance

AI in Regulation and Governance

As AI reshapes how we interact with the world, it will become important to consider new laws that account for the ways in which it disrupts societal norms and ethical boundaries. Let us imagine a scenario where a self-driving vehicle gets into an accident. Without comprehensive regulations, how do we determine accountability in such a case? Is the manufacturer at fault? Can the owner of the vehicle be held liable? This is just a glimpse of the many complexities surrounding AI and its legal ramifications, making it essential to delve into those complexities in more detail.

Developing Effective AI Legislation

AI's rapid development and integration into critical sectors—healthcare, finance, transportation—introduce unique challenges that existing legal frameworks are often ill-equipped to manage. This gap underscores the necessity for specific AI regulations that govern its development and deployment and address potential risks and ethical implications. Unregulated AI could lead to scenarios where biased decision-making algorithms cause harm or personal data is used without consent, leading to privacy violations. Thus, effective legislation is crucial to foster an environment where innovation can thrive under clear ethical and legal guidelines.

The principles of good AI legislation hinge on transparency, accountability, and fairness. Transparency in AI involves clear documentation of AI systems' decision-making processes, enabling users to understand how and why decisions are made. It is important in sectors like healthcare, where AI diagnoses need to be explainable to patients and medical practitioners. Accountability refers to establishing clear guidelines on who is responsible for the outcomes of AI decisions, a necessary measure to ensure that AI systems are used responsibly. Fairness involves AI systems that do not perpetuate existing biases or create new forms of discrimination.

Implementing these principles into law involves a deep engagement with ethical considerations, balancing innovation with responsibility.

We can analyze current examples from around the world to get an understanding of what future regulation might look like. For instance, the European Union's General Data Protection Regulation (GDPR) sets a precedent for how AI should handle personal data, emphasizing user consent and the right to explanation. It is known to have particularly strict consent requirements. On the other hand, Singapore has adopted a similar but slightly more lenient approach with their Personal Data Protection Act, which protects rights for people and issues penalties for businesses that don't comply, but features lesser penalties and looser consent requirements (Neumetric, n.d.)

Looking ahead, the legislative challenges are manifold. One of those challenges will be the regulation of autonomous weapons, which pose both ethical and security risks. Another pressing issue is the use of AI in government surveillance systems, which could lead to invasions of privacy if not regulated properly. Proactive strategies for legislative adaptation might include the establishment of international AI safety standards and the creation of independent regulatory bodies tasked with overseeing AI development across borders. These measures would address immediate concerns and lay down a framework for addressing future problems, while also aligning AI development with global ethical standards and societal values.

AI and Intellectual Property

The intersection of artificial intelligence and intellectual property (IP) law presents a fascinating conundrum that challenges the very framework of how we view creation and invention in the digital age. As AI systems increasingly perform tasks that traditionally require human intellect, such as composing music, creating art, or even drafting patentable innovations, the standard notions of authorship and invention are being put to the test. These developments raise questions about the adequacy of current IP laws, which were designed in an era long before AI's capabilities were imagined.

One of the most pressing issues is the question of authorship when an AI system creates a work. Traditional IP laws are built on the premise that only humans can be authors or inventors, which grants them certain exclusive rights to their creations. However, AI challenges this notion by producing work that can be indistinguishable from human creations without clear legal precedent on how to handle such cases. If an AI algorithm designs a new architectural structure or writes a piece of music, who, if anyone, owns the copyright? The current legal framework does not recognize AI as a rights holder, which leaves a gap in protection and attribution for AI-generated works.

Analyzing the current IP frameworks reveals that while some regions have begun to address these issues, significant gaps remain. For example, the United States Patent and Trademark Office has faced applications for patents created by AI systems. These cases have forced a reevaluation of what constitutes 'inventorship.' Traditionally, patents require a human inventor, but AI challenges this requirement. In February of 2024, the office maintained the stance that AI could not be recognized as an inventor, primarily because the laws were not designed to accommodate non-human creators (USPTO, 2024). USPTO's decision underscores the need for legislative updates to reflect the evolving landscape of innovation.

Case studies of AI and IP conflicts provide further insight into the complexity of these issues. One notable example is a piece of artwork named Portrait of Edmond Belamy, which sold for $435,000 in an auction. It was created using an open-source generative adversarial network, which utilizes a system of judges and creators to build options based on patterns in specific datasets. The artwork was created by a French art collective named Obvious, but some argue the programmer who created the open-source code used to create the artwork, Robbie Barat, should get credit (Jee, 2018). These discussions have yet to lead to a consensus, reflecting the broader uncertainty and lack of legal clarity in this area.

To navigate these areas, recommendations for the evolution of IP law are essential. One approach could be the introduction of a new category of

IP rights specifically designed for AI-generated creations. This category would provide a limited form of protection that recognizes the role of AI in the creative process without granting full rights akin to those awarded to human creators. Additionally, updating the criteria for patentability to include AI-generated inventions could incentivize innovation while recognizing the unique contributions of AI systems. Policymakers must consider the balance between promoting innovation and preventing the monopolization of algorithms that stifle creativity and competition.

The evolution of IP law will likely be a gradual process, marked by continued debate and adjustment as we collectively try to figure out what makes the most sense. As these discussions progress, it will be crucial for policymakers, legal experts, and technologists to collaborate closely to ensure that the laws governing AI and intellectual property effectively reflect our values and promote a just and innovative society.

Privacy Laws and AI

As AI technologies increasingly infiltrate various aspects of daily life—from personalized shopping recommendations to predictive healthcare diagnostics—privacy can be a major concern. AI systems, by their very nature, rely on vast amounts of data to learn, adapt, and make decisions. This necessity for data can bring about a myriad of different issues, particularly around the practices of data harvesting and the potential for invasive surveillance. AI-driven platforms can track user activities across the web to create detailed profiles, often without explicit consent or awareness of the individuals involved. Moreover, the capabilities of AI to process and analyze this data at unprecedented scales further exacerbate privacy risks, highlighted by the danger of sensitive information being exploited or mishandled.

Current privacy laws, while robust in certain regions, often struggle to keep pace with the rapid advancements in AI technology. Existing regulations like the European Union's General Data Protection Regulation (GDPR) have set benchmarks in terms of addressing user consent, data protection, and the right to privacy (European Commission, n.d.). However, even these comprehensive laws have gaps when dealing with

AI-specific issues, such as the continuous learning aspect of AI systems, which can make it difficult to determine when and how data is being used. Other regions with less stringent privacy protections offer even fewer safeguards, raising questions about the inconsistency of privacy standards globally. This disparity complicates the regulatory environment for international AI applications and poses challenges in protecting individuals' privacy across borders.

Proposed changes to privacy legislation are increasingly recognizing the need to adapt to the realities of AI. For example, the 'privacy by design' principles in the development of AI systems stress integrating privacy considerations directly into the technology development process rather than tacking them on as an afterthought. Legislation may need to introduce stricter requirements for transparency in AI operations, so that users are fully informed about how their data is being used and for what purposes. Another critical area for legislative enhancement is the strengthening of user consent mechanisms, ensuring that they are strong enough to handle the nuances of AI data usage, which often extends beyond the initial scope of the user agreement.

Current privacy laws... often struggle to keep pace with the rapid advancements in AI technology.

Balancing the scales between AI innovation and privacy protection is a delicate task. Stringent privacy laws could stifle AI development; lax regulations could lead to breaches of trust and privacy, undermining public confidence in AI technologies. Lawmakers need to navigate this balance, juggling the evolution of privacy laws with AI advancements. This entails maintaining flexibility in legislative frameworks to adapt to future developments, in addition to continuous dialogue between technologists, legal experts, policymakers, and the public to avoid compromising the foundational values of privacy and individual rights.

Because AI technologies operate across global networks, creating more uniform privacy laws internationally could help establish a common standard that protects privacy while supporting innovation. This global approach would encourage a collaborative environment for addressing the complex array of difficulties AI poses to privacy. Moving forward, the collective efforts of the global community will be paramount in shaping a regulatory environment that both nurtures AI's potential and vigorously defends the privacy rights of individuals. This ongoing process will likely see many iterations of lawmaking, each refining the laws and regulations that came before.

Environmental Governance

Monitoring and Compliance

Artificial intelligence offers tools for both advanced monitoring of ecological data and for complying with stringent environmental regulations. AI systems, equipped with the capacity to process vast amounts of data from sensors and satellites in real-time, are increasingly utilized to monitor critical environmental parameters such as air quality indices (Mahsood & Ahmad, 2021) and water purity levels (Gunda, Gautum & Mitra, 2018). These systems can detect minute changes in environmental conditions, alerting authorities and stakeholders to issues before they escalate into crises. AI-driven analytics can track air quality across different regions, identifying sources of pollution and predicting areas at risk of hazardous air quality levels. This capability allows for swift action, alerting populations to smog warnings or adjusting industrial operations to mitigate pollution output.

Traditional methods of regulatory compliance often involve manual inspections and audits, processes that are not only time-consuming but also prone to human error and limitations in scope. AI is different—it enables continuous surveillance and analysis. AI algorithms can be used to analyze emissions data from industrial plants, ensuring that operations stay within legally mandated environmental standards. These systems can predict violations by learning from historical data, thereby preempting non-compliance and aiding in the proactive management of environmental risks. Furthermore, AI facilitates a dynamic approach to environmental regulation, where policy adjustments can be made in real time based on data-driven insights, creating more effective and adaptable environmental governance.

One example is an AI project from the Environmental Protection Agency (EPA), which ranks the risk that people have been exposed to specific chemicals and identifies the pathway that led to the exposure. The EPA's project is capable of classifying exposure into one of four categories: consumer near-field, dietary, far-field industrial, and far-field pesticide to the general population (EPA, 2022). This classification then aids in identifying areas where environmental regulation may need improvement.

The future applications of AI in environmental governance are vast and varied. One of the most promising areas is the management of renewable energy resources. AI can optimize the operation of wind and solar farms by analyzing weather forecasts and adjusting energy output to maximize efficiency. This can enhance the stability of energy grids by predicting and managing fluctuations in power supply. Additionally, AI could revolutionize waste management systems. Through the integration of AI with the Internet of Things (IoT), waste management processes can become more efficient and less resource-intensive. AI could enable smart sorting systems that automatically classify and recycle waste, reducing human labor and improving recycling rates. AI-driven systems could also monitor waste levels and collection routes in real time, optimizing logistics and reducing the environmental impact of waste collection.

These advancements represent just the tip of the iceberg. AI integration into environmental governance will likely become more profound, offering innovative solutions to some of the most pressing environmental challenges of our time. By harnessing the power of AI, policymakers and environmental managers will become equipped with tools that enhance the effectiveness of environmental protection efforts and pave the way for more sustainable and resilient ecological management practices.

Public Services

The automation of routine tasks such as the processing of tax returns or the issuance of public documents are other ways in which AI can improve our lives. In the future, AI systems could be deployed to handle these functions with a level of precision and speed unattainable by human clerks. This automation could reduce processing times and minimize errors, leading to a more streamlined interaction between citizens and government bodies. AI-enabled systems could one day quickly analyze tax submissions, flag inconsistencies, and process returns, reducing the bureaucratic backlog and improving taxpayer satisfaction. Similarly, in document issuance, AI has the ability to eventually manage applications for passports or driving licenses, verifying details against existing databases and issuing documents without human intervention, thus cutting down wait times and reducing the opportunities for clerical errors.

Beyond streamlining operations, AI's impact extends into making government services more accessible to the public, particularly benefiting people with disabilities. Innovations such as voice-activated systems and AI-powered web accessibility tools can break down barriers that have traditionally made interaction with government services challenging for those with disabilities. Voice-activated AI assistants can help users complete online forms or access information without needing to navigate complex websites. For individuals with visual impairments, AI-driven applications can translate textual information into speech or Braille, while those with hearing impairments can benefit from real-time speech-to-text transcription services provided by AI systems, such as with the Live Transcribe app on iPhone and Android and Otter.ai (Goldstein, 2020).

One of the most anticipated developments is in predictive services for public health and safety. AI systems, through the analysis of vast amounts of health data, have the potential to predict outbreaks of diseases, enabling preemptive health measures that can save thousands of lives. For instance, an AI system could analyze trends in healthcare data to identify the locations and times of flu outbreaks before they occur, allowing for increased vaccinations and public health advisories. Similarly, in public safety, AI could analyze crime data to allocate law enforcement resources more effectively, ideally preventing crimes before they happen.

Another promising area is the development of personalized public information systems. The idea behind these AI-driven systems is to provide people with information tailored to their specific needs or circumstances, accessible through simple conversational interfaces. A personalized public information system could inform citizens about local events, voting changes, or public service disruptions based on their location and preferences, communicated through their preferred medium, be it text, voice, or even video.

AI integration into public services will deepen, driven by the dual imperatives of increasing operational efficiency and enhancing citizen satisfaction. The ongoing developments in AI technology promise to streamline and simplify administrative processes, while making government services more responsive and accessible to all citizens, irrespective of their physical abilities or technological proficiency.

Crisis and Disaster Management

AI is able to analyze real-time data from multiple sources during a crisis, synthesizing information from satellites, sensors, social media, and other communications quickly and accurately to provide emergency responders with a comprehensive view of the situation as it unfolds. This allows for rapid assessments, making it possible to direct resources more effectively and anticipate the next phases of a crisis.

AI's predictive capabilities are invaluable in preemptive crisis management. By analyzing patterns from vast datasets, AI models can forecast the trajectory of hurricanes, the spread of wildfires, or the progression of a flooding event with remarkable accuracy. These predictions are already being used by organizations such as the NOAA, FEMA, United States Geological Survey, NASA, and IBM's The Weather Company (Hyper-Reach, 2023). They enable governments and organizations to prepare more effectively, whether by evacuating areas likely to be hit hardest, pre-positioning relief resources, or informing the public about potential dangers. The ability to anticipate the impact of a disaster before it strikes can dramatically reduce both human and economic losses.

By leveraging AI... we can enhance the efficacy and efficiency of our emergency responses.

AI might play a vital role in coordinating disaster response efforts. In the future, AI algorithms could optimize evacuation routes in real-time, taking into account changing conditions, road closures, and traffic flow to guide residents to safety efficiently. This optimization would help to speed up evacuation processes and alleviate congestion, reducing the chaos that can exacerbate already tense situations. AI systems could predict the needs of affected populations, from medical supplies to food and water, ensuring aid is distributed effectively and equitably—and preventing shortages. Steps in this direction have already started to be taken, as the Pentagon and Carnegie Mellon University collaborated on a project called xView2 in 2019. It's a project that uses machine learning algorithms and satellite imagery to assess damage from earthquakes and create an action plan for repairs and rescuing any civilians in need of help (Ryan-Mosley, 2023).

Moreover, AI contributes to building resilience against future disasters. Post-event analysis conducted by AI helps identify vulnerabilities in infrastructure and response strategies, providing critical feedback that can be used to improve future responses. By learning from each event, AI-enhanced systems evolve to manage disasters more effectively, supporting ongoing efforts to build more resilient communities.

Through these applications, AI can shift how we respond to crises from reactive to proactive management. This shift can both save more lives and preserve infrastructure and resources, reducing the long-term economic impacts of disasters.

The exploration of AI's role in crisis and disaster management illuminates its potential to transform how societies prepare for, respond to, and recover from disasters. By leveraging AI for real-time data analysis, predictive forecasting, and strategic response planning, we can enhance the efficacy and efficiency of our emergency responses. From enhancing AI legislation to managing how AI is used in our government processes, there is a lot to do when it comes to incorporating AI into our society in a way that prioritizes both ethics and the technology itself.

Moving forward, our focus will shift to AI's role in reshaping city structure, managing infrastructure, and enhancing quality of life through improvements in education and community well-being. We will also see how AI can improve engagement through social media and support personal wellness.

CHAPTER 6

AI in Life and Society

AI in Life and Society

AI CAN REDEFINE how we all live and interact with each other and society at large—from street lamps to skyscrapers to improving accessibility and transforming the processes of leading companies around the world, it's time to explore how AI is setting the stage for these types of changes, improving the quality of life for city dwellers and rural folk alike.

AI and the Future of Urban Living

Smart Cities and Infrastructure

In the heart of modern urban landscapes, AI is becoming a cornerstone of development, leading to the creation of smart city ecosystems where technology and data drive better, more efficient city management and services. These smart cities can harness AI to optimize everything from traffic systems to waste management, thereby enhancing the quality of urban living while ensuring sustainability.

Smart City Ecosystems

Smart city ecosystems utilize AI to integrate and synthesize data from various sources, including sensors, cameras, and IoT devices, across the city. This integration facilitates the automation and optimization of numerous public services. AI-driven public transport networks can adjust routes and schedules in real time based on traffic conditions and rider demand, minimizing wait times and congestion. Similarly, smart energy systems use AI to predict peak demand periods and adjust supply accordingly, enhancing energy efficiency and reducing costs.

AI can play a pivotal role in public safety within these ecosystems. Through the analysis of data from surveillance cameras and other sensors, AI can help predict and mitigate potential public safety threats, from identifying suspicious activities to optimizing the deployment of emergency services during incidents.

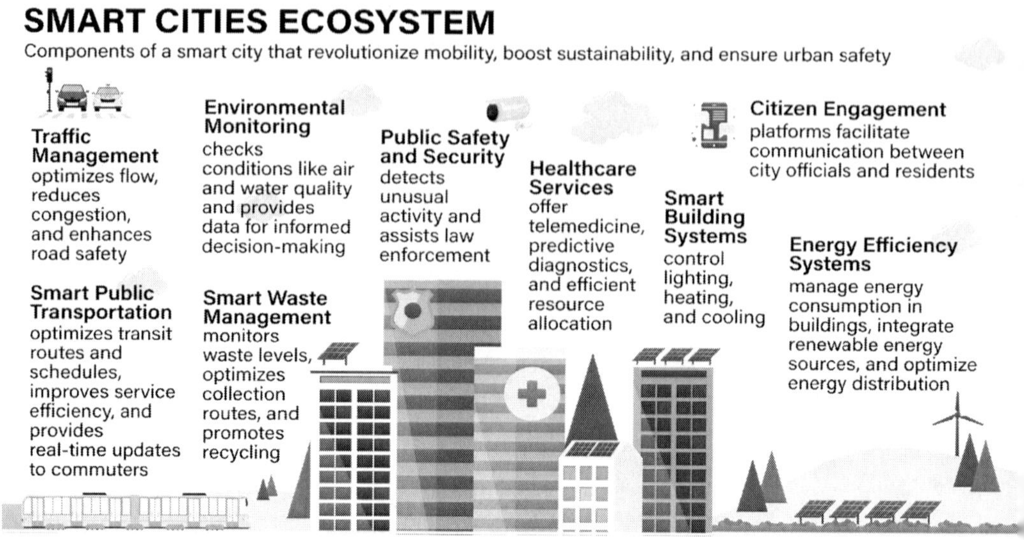

Infrastructure Management

AI's integration into urban infrastructure management can revolutionize how cities handle maintenance and development. With AI-powered predictive maintenance, cities can anticipate and address infrastructure issues before they escalate into serious problems. For example, AI algorithms can analyze data from sensors embedded in roads and bridges to predict structural failures or deteriorations, allowing for timely repairs that prevent costly and disruptive breakdowns.

This predictive capability can extend to the management of urban water systems, where AI models predict pipe failures and leakage points, keeping an uninterrupted and efficient water supply while conserving a vital resource. As a result, the costs of maintenance are reduced and the safety and reliability of urban infrastructure is improved.

Environmental Monitoring

As discussed earlier, AI systems can analyze environmental data from various sensors to monitor air quality, water quality, and pollution levels in real-time. However, AI-enabled environmental monitoring can aid in urban planning. By analyzing patterns and predicting future environmental conditions, such as heat islands or flood-prone areas, city planners can design urban spaces that are both resilient and sustainable. This could be a big key to addressing the challenges posed by climate change and urban density.

Quality of Life Improvements

Perhaps most importantly, AI-driven innovations in smart cities are poised to dramatically enhance the quality of life for residents. Personalized public services powered by AI will customize health and social services to individual needs, improving access and effectiveness. In education, AI systems are expected to provide insights that align public educational programs with job market trends, enhancing employment opportunities for residents. Additionally, smart city apps and platforms will enable residents to interact directly with city management systems, reporting issues and providing feedback.

The integration of AI within urban ecosystems is set to redefine the essence of urban living. Smart cities equipped with AI-driven systems promise a more sustainable and efficient urban future, alongside improved quality of life for residents. As these technologies continue to evolve and scale, the potential for transformative change in our urban environments is boundless.

Social Media and AI

Transforming Engagement and Content Creation

In the digital age, your experience on social media is largely shaped by sophisticated AI algorithms, whether you know it or not. These algorithms curate and personalize every aspect of your content feed. They analyze myriad data points: what you like and share, the time you spend on different posts, and even the nuances of your interactions. This data is then used to tailor your social media feeds in real, innately personal ways, ensuring that the content you see is uniquely suited to your preferences and likely to capture your interest. Personalization extends beyond just showing you posts from your friends; it may include targeted advertisements and news content that align with your behaviors and expressed preferences.

Whether it's YouTube, Facebook, Instagram, or even LinkedIn, each platform has its own algorithms. YouTube takes note of what videos you've watched in the past, videos that are commonly watched together, and even what topics you watch the most videos about. Facebook takes note of what you tend to interact with and the content type you like most. Instagram features multiple algorithms—one for reels, one for the feed, one for stories, and one for the explore page (Newberry, 2024). What they have in common, though, is that they're all curating content for you.

For businesses and content creators, this AI-driven curation is a double-edged sword. On the one hand, it enables them to reach audiences who are most likely to find their content appealing, increasing engagement rates. On the other hand, it means navigating an ever-changing algorithm to ensure their content remains visible to their audience. It requires a deep understanding of how AI works and continually adapting strategies to leverage these AI systems effectively.

AI's influence can actually extend further into optimizing user engagement through the strategic timing and formatting of posts. By analyzing large datasets on user activity, AI can predict when specific users are most likely to be online and engaged, suggesting optimal times for posting to maximize visibility and interaction. This is invaluable for marketers and social media professionals who seek to increase the impact of their campaigns. Furthermore, AI tools can analyze performance data from past posts to determine which types of content formats (videos, images, long-form articles, etc.) perform best with particular demographics, enabling content creators to tailor their production strategies to audience preferences effectively.

The creation of digital content itself has also been revolutionized by AI. Tools powered by AI are now capable of generating written content, enhancing images, and even creating video effects that would have required significant human effort and expertise in the past. These tools now allow individuals and businesses to produce high-quality digital content without the need for extensive resources. AI-driven graphic design tools can automatically generate stunning visuals based on basic input parameters, while AI video editors can quickly cut and assemble video clips based on predetermined styles or themes.

Addressing Misinformation

The power of AI in social media is not without any challenges. One of the most critical issues is the role of AI in the spread and management of misinformation. AI algorithms designed to maximize engagement can inadvertently promote sensational or misleading content that is more likely to generate clicks and shares. Combatting this issue will probably require sophisticated AI-driven tools that can detect and flag false information, distinguishing between credible and questionable sources. However, the nuances of human language and the subtleties of misinformation make such a thing difficult.

Current AI technologies struggle with understanding context and the complex web of human emotions and biases that often characterize misleading information. While AI can be trained to identify blatantly false claims based on data patterns and discrepancies, subtler forms of misinformation, such as those involving irony, satire, or deep cultural contexts, often evade AI detection. This limitation necessitates ongoing research and development to enhance the sophistication of AI models, in addition to a collaborative approach involving human oversight to effectively manage and mitigate misinformation on social media platforms.

The Role of AI in Personal Wellbeing

In the area of personal health and fitness, artificial intelligence can offer tools and insights that enhance physical wellness, elevate mental health, and improve overall quality of life. AI is capable of reshaping how individuals manage their health routines, adhere to personalized diet plans, and maintain mental well-being.

AI-driven health platforms are increasingly capable of analyzing vast amounts of data from wearable devices that monitor everything from heart rate and sleep patterns to blood glucose levels and physical activity. This data, when processed through sophisticated AI algorithms, enables the creation of highly tailored health and fitness programs. For instance, AI can suggest personalized workout plans that adapt in real-time based on your performance and progress or dietary recommendations aligned with your nutritional needs and fitness goals, taking into account past eating habits and even genetic predispositions.

> *...AI can create a shift in how we balance our professional and personal lives.*

AI's impact even extends into mental health management, where it aids in the detection and treatment of issues such as depression and anxiety. By analyzing speech patterns, facial expressions, and behavioral data collected through apps and devices, AI can identify signs of mental distress early. It can offer support through therapeutic tools and programs that provide coping mechanisms and track your progress over time. This proactive approach helps in early intervention and reduces the stigma associated with seeking mental health care by making it more accessible and less intrusive.

Beyond individual health management, AI can create a shift in how we balance our professional and personal lives. AI-powered tools enable smarter scheduling and workload management. These tools can use AI to analyze work patterns and predict periods of high stress or burnout, suggesting optimal breaks and moderating workloads. For remote workers and freelancers, AI-driven apps can optimize project timelines and automate routine tasks, freeing up time that can then be spent with family or on leisure activities, enhancing overall life satisfaction.

AI and Accessibility

AI technologies are breaking down barriers that traditionally isolated people with disabilities. For example, AI-powered assistive devices can convert spoken language into text or sign language in real-time, helping those with hearing impairments better interact in their daily lives. Slait.ai is one such tool currently in the works. It's an app that converts sign language to text, speech to text, and even has tools to enable video communication with these functionalities (Slait.ai, n.d.). Visual recognition AI can assist visually impaired individuals by describing their surroundings or reading text from books and screens, thus enabling greater independence.

In the home, AI is being integrated into smart technologies that assist individuals with mobility problems. These smart systems can control lighting, temperature, and even door locks via voice commands or smartphone apps, enhancing the ability to live independently (Cericola, 2024).

AI and Education

Transforming Learning Environments

In education, AI can personalize content to meet the diverse needs of students. This involves analyzing individual learning patterns, preferences, and speeds to tailor educational content accordingly. Such systems dynamically adjust the difficulty of tasks, the pace of lessons, and the style of content presentation to optimize each student's learning experience. If a student excels in visual learning but struggles with textual information, AI can modify the curriculum to present more graphical or video-based content, thereby enhancing the student's ability to grasp and retain complex concepts.

This personalization extends into the burgeoning field of AI tutors and assistants, which provide additional support outside the traditional classroom setting. These AI-driven systems are designed to mimic the one-on-one interaction students might have with a tutor, offering explanations, supplementing learning materials, and providing feedback in real time. Technologies like the Carnegie Cognitive Tutor have been used to help with high school mathematics, and apps like Duolingo and Carnegie Speech can help with foreign languages. Advanced tutors for specialized subjects are starting to pop up as well, such as SHERLOCK, which teaches Air Force technicians to diagnose electrical problems (Stone et al., 2016). What makes these AI tutors particularly effective is their ability to be available 24/7, removing barriers to learning that can arise from scheduling conflicts or the lack of availability of human tutors. AI's constant availability provides students with help the moment they encounter a hurdle in their learning process, thus maintaining their momentum and engagement.

AI'S ROLE IN ENHANCING EDUCATION

AI used in education through personalized learning, streamlined administration, and immersive virtual classrooms.

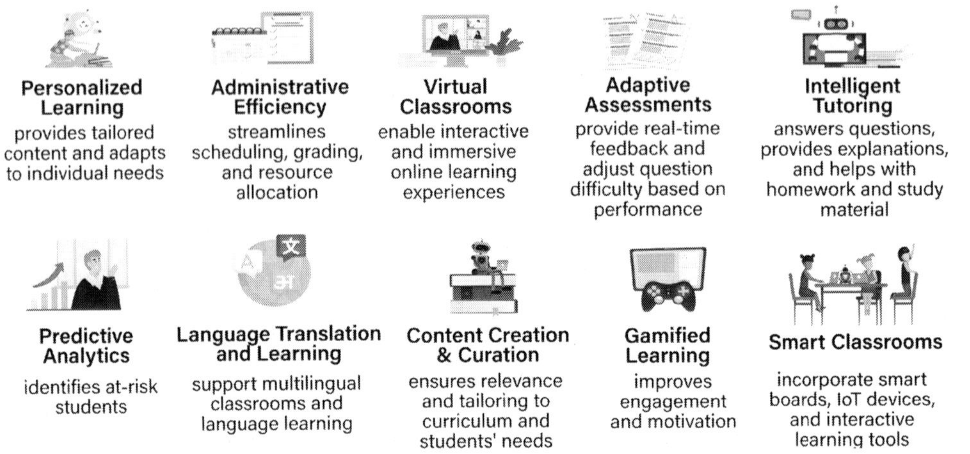

AI's role in transforming educational environments can also include enhancing the accuracy and efficiency of assessments. Traditional methods of evaluating student performance are often time-consuming and subject to human error and biases. AI-driven assessment tools can revolutionize this process by automating the grading of standardized tests and providing more nuanced analytics on student performance. These tools can analyze not just what answers students get wrong but how they approach problems, providing insights into their thinking processes and areas where they may need more support. This allows educators to provide targeted interventions and to better understand the effectiveness of their teaching strategies.

Operational Efficiency in Educational Institutions

AI can boost operational efficiency within educational institutions by automating administrative tasks such as attendance, grading, and scheduling. It can free up educators to focus more on teaching and spending time with students and less on the bureaucratic aspects of education. Furthermore, AI-driven analytics can help school administrators identify patterns and trends in student enrollment and performance, aiding in more informed decision-making regarding curriculum changes, resource allocation, and institutional policies.

Looking toward mid-century, AI's impact in education should extend to creating fully personalized learning environments where each student's educational journey is tailored precisely to their needs and goals. AI systems might adapt to individual learning styles or anticipate learning challenges and provide preemptive support. Meanwhile, real-time analytics could continuously refine learning models, making education a truly adaptive and responsive experience.

How Leading Companies Use AI

Google, Facebook, and Amazon

In an era where artificial intelligence is just emerging, leading companies like Google, Facebook, and Amazon are at the forefront, harnessing AI to redefine how search, social media, and e-commerce work. This section covers how these giants leverage AI to reshape our interactions with digital platforms, making personalized experiences the norm rather than the exception.

Google: Search and Information Retrieval

Google, a pioneer in search engine technology, has integrated AI to transform the very essence of search and information retrieval. By employing sophisticated AI algorithms, Google has enhanced the accuracy and relevance of search results, tailoring them to individual user preferences and contexts. This personalization is achieved by analyzing vast amounts of data on user behavior, including past searches, browsing history, and geographical location, allowing Google to anticipate user needs and provide search results that are accurate and aligned with the user's specific intent. Google's AI algorithms continuously learn from new data, adapting and improving over time, which ensures the search engine evolves in sync with changes in language use, internet trends, and collective human knowledge. In Google's October 2020 update to their blog, they shared several ways in which they use AI in their search functionalities. These include the use of spelling algorithms that utilize a neural net to decipher misspellings, an algorithm to better rank the relevancy of specific passages within a search, and neural nets to better understand subtopics around an interest (Raghavan, 2020). Google AI strategies include the following:

- **Advertising and Marketing:** Through machine learning models, Google's advertising platform can analyze patterns in user data to identify promising customer segments and predict which ads are most likely to resonate with different audiences. This capability allows advertisers to target their campaigns more precisely, which increases conversion rates and maximizes return on investment. The use of AI in optimizing ad placement and bidding strategies further helps to keep marketing resources allocated in the most economically efficient manner.

- **AI in Products and Services:** Google's integration of AI extends beyond search and advertising into a plethora of products and services that enhance daily life. Google Assistant, an AI-powered virtual assistant, uses natural language processing and machine learning to understand and respond to user queries with high accuracy. This assistant makes everyday tasks, from setting reminders to retrieving information, simpler and more interactive. In Google Photos, AI-driven image recognition technology automatically categorizes photos based on content, making it easier for users to search and organize their memories. Waymo, Google's autonomous car project, epitomizes the application of AI in developing self-driving technology that promises to revolutionize transportation by improving safety and efficiency on the roads.

Facebook: Content Personalization

Facebook's use of AI in personalizing content is a testament to its commitment to enhancing user engagement. By analyzing data points such as the posts users interact with, the time they spend on different types of content, and the networks they are part of, AI algorithms curate a News Feed uniquely tailored to each user. Facebook's algorithms use a tool that it developed called DeepText to understand the meaning behind words that are posted and analyze them contextually. They do this through a form of unsupervised learning, which helps them not get tripped up on spelling, slang, or syntax. With an understanding of the text, they then direct people toward products or content relevant to their posts and conversations, thereby enhancing the overall stickiness of Facebook's services (Marr, 2016). Facebook uses these AI strategies:

- **Advertisement Targeting:** By leveraging AI to analyze user demographics, interests, and behaviors, Facebook provides advertisers with powerful tools to reach their target audience with precision. Facebook uses deep neural networks to make decisions on which advertisements should be shown to which users (Marr, 2016). This targeted approach ensures users see ads that are relevant to their needs and preferences, enhancing the user experience while providing advertisers with higher conversion rates.
- **Content Moderation:** Facebook employs AI to maintain a safe and respectful online environment. AI algorithms are trained to detect and remove content that violates community standards, including hate speech, misinformation, and harmful content. This automated system allows for the rapid and scalable moderation of posts, which is crucial given the vast amounts of content generated by users daily.
- **Face Recognition:** Facebook's implementation of AI extends to face recognition technology used in photo tagging, which automatically recognizes and suggests tags for people in photos uploaded to the platform. It utilizes an application called DeepFace. This feature enhances user convenience and prompts users to engage in social interactions and content sharing. However, it raises important privacy considerations too, so it has faced backlash from EU legislators (Marr, 2016).

Amazon: Recommendation Systems

Amazon's AI-driven recommendation systems enhance consumer shopping experiences. By analyzing past purchases, browsing histories, and product search behavior, Amazon's algorithms can suggest products users are likely to be interested in, creating a personalized shopping experience that encourages discovery and increases customer satisfaction. This system makes shopping more intuitive and less time-consuming while boosting sales by cross-selling and upselling products based on user preferences. Amazon recommendation systems use AI technologies such as:

- **Supply Chain Optimization:** In supply chain management, Amazon leverages AI to predict consumer demand, evaluate the availability of products, optimize delivery routes, track activity on the supply chain, and personalize interactions (Harrigan Jr., 2022). As a result, Amazon can minimize shipping times and costs while making products available when and where they are needed, enhancing customer satisfaction and operational efficiency.

- **Customer Service:** Amazon's customer service greatly benefits from AI through the use of chatbots and voice-activated assistants like Alexa. Alexa even now enables voice shopping, which means customers can search for products, add them to their cart, and check out without needing to touch their device (Harrigan Jr., 2022). AI-powered tools can handle routine customer inquiries and support requests, providing quick and accurate responses around the clock. This automation improves response times and allows human customer service representatives to focus on solving more complex issues, enhancing overall service quality.

- **Automated Retail Stores:** Amazon has begun to use AI to automate a chain of retail stores called Amazon Go. These stores make shopping more convenient, as customers don't have to wait in line or even check out to purchase their items. In the Amazon Go stores, Computer Vision technology monitors when items are removed or replaced on a shelf, then it keeps track of which items you have taken in a virtual cart, and when you leave the store, it automatically charges your Amazon account (Harrigan Jr., 2022).

As these examples illustrate, Google, Facebook, and Amazon are redefining the possibilities within their respective domains, setting benchmarks for innovation, efficiency, and personalization. These advancements are only some examples of the transformative power of AI in shaping the future of technology and society.

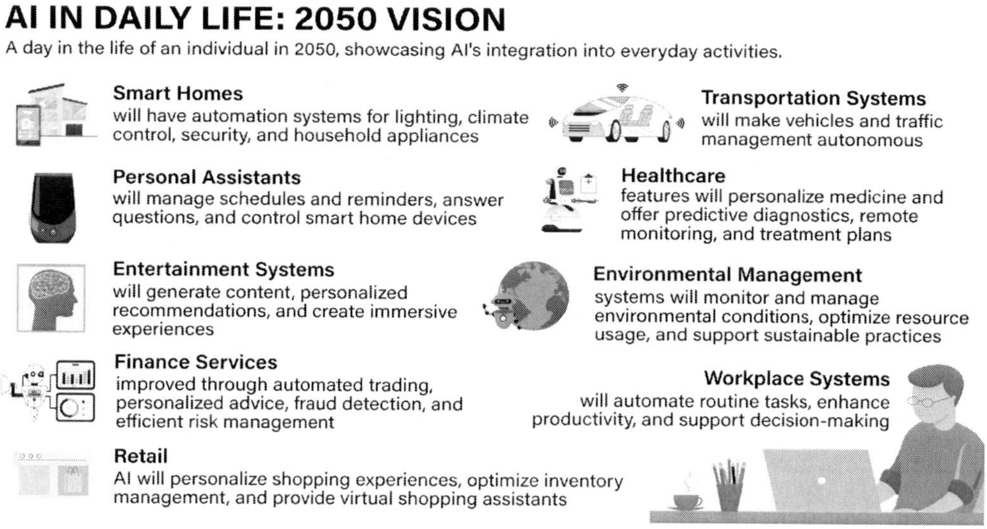

AI's Influence on Daily Life by 2050

The integration of Artificial Intelligence into our homes promises a shift toward more intuitive living environments. This means AI could be deeply embedded in everyday appliances, optimizing everything from your morning routine to energy consumption. And it has the potential to create spaces that actively contribute to your well-being and adapt to your changing needs without explicit commands.

Predictive Home Assistance

The concept of predictive home assistance revolves around AI's ability to anticipate and act upon the needs of household occupants. In the future, smart appliances equipped with AI should respond to direct commands and anticipate needs based on usage patterns and preferences. For instance, a refrigerator could monitor its contents, suggest recipes based on the available ingredients, and even order groceries as items run low. Similarly, an AI-integrated HVAC system could learn your schedule and temperature preferences, adjusting the climate within your home to optimize comfort and energy usage before you even step through the door.

The real power of predictive home assistance lies in its potential to reduce energy consumption and enhance sustainability. AI systems can dynamically adjust energy usage based on peak load times, weather conditions, and real-time energy supply fluctuations, such as those provided by renewable sources. This can both reduce the carbon footprint of your household and aid in managing expenses associated with energy consumption.

Communal and Social Interactions

Beyond the confines of individual homes, AI may also one day transform communal and social interactions, revolutionizing the way communities engage with each other and manage resources. For example, AI-driven platforms can facilitate more effective community governance by analyzing data on community needs and resource usage, thereby aiding in decision-making processes. These platforms will then foster a new level of social interaction by connecting individuals with shared interests or facilitating community-wide events based on collective preferences.

> *...AI may also one day transform communal and social interactions, revolutionizing the way communities engage with each other and manage resources.*

Additionally, AI shows promise in the area of analyzing communication patterns and emotional cues to aid in conflict resolution. AI systems might in the future mediate disputes and offer solutions that conform to previously agreed-upon community standards. This is particularly promising in diverse urban settings, where cultural and linguistic divides need to be navigated with care.

Similarly, AI can facilitate your interactions with others, suggesting activities based on your interests and social circles and even helping you plan gatherings with friends or community events. As you interact with your neighbors, AI-mediated platforms could help to address everyone's needs, promoting a harmonious living environment.

Deep integration of AI in daily life will raise challenges, such as privacy issues and the need for comprehensive cybersecurity measures.

However, we can overcome these issues to create a future of enhanced convenience, improved quality of life, and a stronger sense of community.

From Social Media to the way we buy things and the way we learn and take care of our health, AI touches us already in all aspects of our lives. The developments just discussed showcase broad implications for society. Going forward, we will dive deeper into the nature of investment and the market when it comes to AI, building on our earlier exploration of startups and why it is important.

CHAPTER 7

Investment and Market Trends

Investment and Market Trends

AI'S INFLUENCE ON investment trends and market dynamics is profoundly reshaping economics. By 2050, we may expect that it will have transformed the stock market with tools such as predictive analytics making the financial industry a highly efficient, data-driven ecosystem. Investment strategies are already no longer just about intuition and historical data but about real-time insights and forecasts powered by AI. This fusion of AI and investment will only continue.

Identifying AI Investment Opportunities

When considering AI investments, you're going to have to find a way through all of the noise to pinpoint technologies that are innovative and poised for growth. This process begins with an observation of technological advancements and market demands. For instance, let's consider the rise of natural language processing (NLP) technologies. As businesses seek to enhance customer interaction through more sophisticated AI-driven communication tools, investments in startups specializing in NLP technologies have seen a large uptick.

Evaluating the viability of these AI startups requires a meticulous analysis of several key factors. Innovation uniqueness, or the distinctiveness of the technology, is especially important. A startup that offers a novel solution to a prevalent market problem, such as AI-driven cybersecurity defenses that adapt to new threats in real time, will stand out as a compelling investment. On the other hand, the strength of the startup team cannot be overlooked. A team possessing both deep technical expertise and a strong understanding of the market landscape is more likely to successfully navigate the challenges of scaling an AI solution.

The sector-specific opportunities in AI investment are vast and varied. In healthcare, AI is already revolutionizing patient diagnosis and treatment. Companies that leverage AI to offer more accurate diagnostics and personalized medicine are reshaping healthcare delivery and are likely a good investment. Similarly, in finance, AI's ability to perform complex quantitative analyses and manage large volumes of data is transforming asset management and fraud detection processes. Automotive is another sector where AI is making significant inroads, particularly in the development of autonomous vehicles and advanced driver-assistance systems (ADAS).

Investment opportunities in AI can broadly be categorized into long-term and short-term opportunities. Long-term investments often focus on foundational AI infrastructure—such as advanced neural networks and computing hardware that supports the development of AI applications. These investments are crucial for the sustained growth and integration of AI technologies across various sectors. On the other hand, short-term opportunities might focus on specific AI applications that address immediate market needs, like AI-powered e-commerce recommendation systems designed to enhance online shopping experiences.

AI's impact on traditional industries offers a dual perspective: disruption and transformation. Sectors like manufacturing and retail are undergoing large changes as AI automates complex processes, enhances supply chain efficiency, and provides deeper consumer insights.

Risk Assessment in AI Ventures

Understanding the risks inherent in AI investments is about making strategic decisions that align with both current technological trends and future developments. The primary risks include technological obsolescence, regulatory changes, and market acceptance. Technological obsolescence is a big concern in the rapidly evolving world of AI, where today's cutting-edge technology quickly becomes outdated. Regulatory changes pose another risk, as government policies adapting to new AI applications can impact the operational feasibility and profitability of AI ventures. Lastly, market acceptance is crucial; no matter how innovative a technology is, its success ultimately depends on people willing and eager to use it.

Mitigation of these risks begins with diversification. Investing across different AI domains and sectors can protect against losses in any single area. While you might invest in AI-driven healthcare diagnostics, diversifying into AI applications in financial services or customer experience management can balance the risks across different market dynamics. In-depth due diligence is another essential strategy. This involves assessing the financial health of the AI venture, in addition to understanding the technology's scalability, the business model's adaptability, and the founding team's vision and expertise. Ultimately, staying informed about technological and regulatory trends is crucial if you want to anticipate changes that could impact your investments. Attending industry conferences, participating in tech forums, and subscribing to leading AI research publications can provide insights into emerging trends and regulatory shifts.

Evaluating a company's readiness to adopt and benefit from AI technologies is a critical aspect of risk assessment. This evaluation should focus on the company's existing technological infrastructure, the technical acumen of its team, and its strategic vision for incorporating AI. For example, a manufacturing company that has already invested in automated and data-driven processes may be better positioned to integrate AI into its operations than a company with outdated systems and processes. Assessing AI readiness not only helps in identifying ventures with the potential for successful AI integration but also highlights areas where investments in training or infrastructure might be required to realize the full benefits of AI.

The investment horizon—long-term versus short-term—plays a crucial role in risk assessment. Long-term investments may require long periods of time to mature but could also redefine industries. These investments carry inherent risks associated with early-stage technologies, including higher rates of failure and longer periods before achieving profitability. Conversely, short-term AI investments might offer quicker returns but can be susceptible to rapid changes in market conditions or consumer preferences. Therefore, understanding and aligning the investment horizon with your risk tolerance and investment strategy is crucial in navigating the complexities of AI ventures.

Navigating the investment landscape in AI requires a nuanced understanding of both the opportunities and the risks involved. By employing robust risk assessment strategies and maintaining an agile approach to investment, you can mitigate losses and position yourself to capitalize on the potential of AI.

The Future of AI in Venture Capital

As the terrain of venture capital continually evolves, AI is likely to be a big point of interest and opportunity. Observing the current trends in venture capital funding for AI startups provides us with a snapshot of a dynamic marketplace where deep learning, robotics, and AI-driven analytics are attracting significant financial inflows. These technologies have profound implications for productivity and efficiency across various sectors. As we've seen throughout the previous chapters, they are already delivering tangible impacts, drawing the attention of savvy investors and venture capital firms globally. Data published by OECD Data in the World Economic Forum shows that the USA and China are leading the charge with 290 billion and 120 billion dollars, respectively, invested in AI in the last 5 years (Gupta, Chopra, & Kasare, 2024).

The role of corporate venture capital (CVC) in the AI space is particularly noteworthy. In recent years, there has been a perceptible shift as more corporations recognize the strategic value of investing in AI startups. This involvement goes beyond mere financial interests; it is fundamentally about fostering innovation that aligns with the corporation's long-term strategic goals. For instance, a tech conglomerate might invest in AI startups that are pioneering advancements in machine learning algorithms to strengthen their ecosystem of AI capabilities. The implications here suggest a future where corporate venture capital could become a critical driver of AI innovation, reshaping industries and redefining competitive landscapes.

In the future, the focus is likely to intensify around AI technologies that promise to revolutionize sectors such as healthcare and automotive. The data from the OECD reveals that AI applications in mobility and autonomous vehicles have seen the most funding in both the US and

China in the last five years, while healthcare is second in the US and fourth in China just behind Robot Sensors/IT Hardware and Business Processes (Gupta et al., 2024).

The implications here suggest a future where corporate venture capital could become a critical driver of AI innovation.

Geographically, we are beginning to see emerging hotspots for AI development beyond the traditional tech hubs like Silicon Valley. Cities around the world, from Bangalore to Tel Aviv, are nurturing vibrant tech ecosystems where AI startups flourish, supported by academic institutions, government funding, and a growing network of venture capital. Startup Genome's annual ranking of the global startup ecosystem in 2024 supports this, listing 10 international cities in the top 15, including London coming in tied with NYC for second overall (Startup Genome, 2024), These international hotspots are likely to attract more investments as they continue being cradles for innovation and technological advancements.

The impact of economic cycles on AI investments adds another layer of complexity to this dynamic market. Like any sector, AI investments are not immune to the broader economic conditions that influence venture capital markets. During economic downturns, investment in riskier assets such as AI startups might wane as investors seek safer havens. However, the inherent resilience and long-term potential of AI ventures, coupled with accelerating digital transformation across industries, suggest that AI investments might recover and even thrive faster than other sectors. To navigate these cycles effectively, investors should adopt a balanced approach and stay informed about technological trends and market dynamics.

As the AI landscape continues to evolve, so does the venture capital ecosystem that fuels its growth. The strategic integration of corporate venture capital, the emergence of global hotspots for AI development around the world, and the adept navigation of economic cycles are shaping a future where AI investments are poised to change the world. You can be a part of that.

AI and Consumer Behavior

Predicting the Trends

Understanding how consumers make decisions and what influences their behaviors is a cornerstone of modern marketing. As artificial intelligence integrates deeper into this domain, it leverages big data to unlock patterns and trends previously imperceptible to human analysts. Through such a process, businesses are able to understand consumer behavior better and predict future actions with an increased degree of precision. Consider an AI system analyzing millions of transactions to identify buying patterns that correlate with specific demographics, times of year, or even changes in the weather. This data, when processed through advanced AI algorithms, provides businesses with actionable insights that can drive marketing strategies and product development with an unprecedented focus on consumer needs and preferences.

AI-driven recommendations and augmented reality (AR) shopping experiences can redefine how consumers interact with products before making a purchase decision. AI algorithms suggest products that a consumer is likely to be interested in based on sophisticated pattern recognition analysis that includes previous purchases, items viewed, social media interactions, and even behaviors of similar users. Augmented reality takes this a step further by allowing consumers to visualize products in their own space before buying—be it clothing on their avatar or furniture in their living room. Big companies like Ikea and Toyota are already taking advantage of strategies like these for their brands, while smaller companies like PacSun, a clothing brand, and Vanity Planet, a skincare company, have made strides to incorporate AI for the benefit of the customer (Klavins, 2023). This makes shopping more interactive and personalized while reducing the likelihood of returns, a significant cost for online retailers.

The interaction between AI and consumers is likely to evolve in ways that will further embed technology into daily buying activities. As AI technologies become more sophisticated, there is potential for a shift in consumer privacy expectations and behavior. The increasing capacity of AI to analyze personal data raises valid concerns about privacy and data security. Consumers are becoming more aware of the value and sensitivity of their personal information. In response, we are likely to see more demand for transparency in how companies collect, use, and share consumer data. Additionally, the regulatory landscape around consumer data will continue to evolve, probably affecting how businesses deploy AI in consumer interactions.

Integration of AI in consumer behavior analysis should lead to the development of even more advanced predictive models. These models could anticipate needs and offer solutions before the consumer even recognizes a desire for something and the potential for creating truly personalized experiences is enormous. This evolving dynamic will require businesses to balance innovation with ethical considerations, ensuring that the use of AI in understanding and predicting consumer behavior adheres to evolving norms surrounding privacy and trust.

AI in Real Estate

Predictive Analytics and Market Insights

AI can reshape how property values are assessed, maintenance schedules are determined, and investment decisions are made. Traditional methods are often reliant on manual comparisons and somewhat subjective interpretations. On the other hand, AI systems have the ability to aggregate and analyze data with precision at unprecedented scales. AI algorithms can swiftly analyze changes in property prices within a specific locale, adjusting valuations in real time to reflect current market conditions. This capability increases the accuracy of property valuations and enhances the transparency and speed of real estate transactions, benefiting buyers, sellers, and investors alike (RealSpace, 2023).

Predictive maintenance, supported by AI, may play a significant role in property management. Traditional approaches to maintenance typically involve scheduled checks or responding to issues as they arise. AI will introduce a new, proactive strategy by utilizing data from various sensors installed in buildings to predict when a component might fail or when a system is likely to need servicing. This method avoids the inconvenience and costs associated with unexpected breakdowns while extending the lifespan of the property's infrastructure. For property managers and owners, the strategy means enhanced efficiency, reduced operational costs, and improved value of their investments. This would be especially useful in the commercial real estate sector, where the scale of operations often makes traditional maintenance methods cumbersome and less effective.

This improves the operational efficiency of commercial properties and contributes to cost savings and sustainability by reducing energy consumption.

In the domain of real estate investment, AI can use sophisticated decision-support tools that provide investors with detailed market forecasts and risk assessments. These tools analyze market trends, real estate cycles, and demographic data to offer predictions about property market behaviors. For investors, this means access to insights that can inform smarter, data-driven decisions. Whether it's identifying emerging markets with potential for growth or assessing the risk profile of a particular real estate investment trust (REIT), AI equips this group with the tools to navigate the complexities of the real estate market with greater confidence.

The impact of AI on commercial real estate encompasses the optimization of space utilization and energy management—critical aspects in the management of commercial properties. AI systems can monitor and analyze how different spaces within buildings are used, providing data that can help in optimizing everything from heating and lighting to the allocation of office spaces. This improves the operational efficiency of commercial properties and contributes to cost savings and sustainability by reducing energy consumption.

Market Trends and AI's Expanding Influence

AI technology is no longer confined to tech-centric sectors. As we've seen in previous chapters, it's spreading its tendrils into realms like agriculture, education, and public services, adjusting the pace and approach of operations in these fields. The result is that AI is influencing market dynamics in a profound way. Consumer expectations have been recalibrated. Users now expect highly personalized experiences across services and products, a demand that is driving companies to adopt AI to create more customized offerings.

Streaming services like Netflix and Spotify use AI to analyze user preferences and viewing habits, providing personalized content recommendations that enhance user engagement and satisfaction. Similarly, AI's impact on competition across industries is pushing businesses to innovate relentlessly, often transforming traditional business models. Retail giants, for instance, are leveraging AI to enhance customer experiences and optimize supply chains and logistics, driving efficiency and reducing operational costs.

Emerging markets, with their rapid development and adoption of mobile technology, present a unique opportunity for AI expansion. These markets are increasingly harnessing AI to leapfrog traditional stages of technology adoption, directly implementing advanced AI solutions in areas such as mobile banking, online education, and digital health services. However, they also bring challenges, including the need for digital infrastructure and stringent data protection laws to manage the increased data flow and ensure user privacy and security.

Speculating on future market trends influenced by AI, we can imagine the emergence of entirely new markets and business models. The rise of autonomous vehicles could create a market for AI-driven logistics and fleet management services. Similarly, AI's advancements in language processing might foster new enterprises offering real-time translation services, eliminating language barriers and opening avenues for global interaction. These developments highlight AI's potential in creating new markets and its role in driving economic changes and opportunities across the globe.

We have explored the expansive role of AI in reshaping market trends and investments across various industries, from transforming investment strategies and consumer expectations to enabling the rise of new markets in emerging economies. AI's influence is both game-changing and expansive. The time has come to shift focus by addressing key challenges and concerns surrounding AI, diving deeper into how we can tackle things such as changes in the job market, the security vulnerabilities surrounding AI's vast datasets, bias problems, and existential risks.

CHAPTER 8

Balancing Challenges and Concerns

Balancing Challenges and Concerns

WITH AI BECOMING more integrated into our society and our lives, we should, and hopefully will, turn a discerning eye toward the challenges it introduces. These issues range from ethical issues, such as bias and fairness in AI decision-making, to technical problems, like ensuring the security and robustness of AI systems, or the impact of AI on employment. Let's delve into some of these areas and offer insights into how we can tackle them.

AI and Cybersecurity

With AI systems becoming more embedded in our networks and devices, they also become targets for new forms of cyberattacks. The sophistication of AI can be a double-edged sword; the very intelligence that allows AI systems to predict and adapt to user behavior can be exploited to create more cunning AI-driven cyberattacks. For instance, AI can be used to automate the creation of phishing emails indistinguishable from legitimate communications or to develop malware that adapts in real time to the defenses it encounters.

However, AI can also be used to combat those kinds of cyberattacks. AI systems are capable of monitoring vast networks to detect anomalies that may elude human oversight. They can process and analyze mountains of data in real time, identifying threats before they manifest into breaches. AI-driven behavioral analysis tools can detect unusual patterns in network traffic, identifying and thwarting cyber threats before they compromise data (Sec1, n.d.). These tools learn from each interaction, continuously improving their detection algorithms.

In the event of a cyberattack, AI systems have the ability to respond autonomously. For example, if they detect an infected device, they can quarantine it, and if they discover a malicious IP address trying to access the network, they can block access in real time (Sec1, n.d.)

CHALLENGES OF IMPLEMENTING AI

Challenges faced by various industries in implementing AI, such as cybersecurity, ethical concerns, and technical limitations.

Data Privacy and Security
Organizations must protect sensitive information from breaches and misuse

AI Transparency and Explainability
Ensuring AI systems are transparent and their decisions are explainable to users is a critical challenge.

Digital Divide
Ensuring equitable access to AI technologies and resources is a significant challenge

Workforce Transition
Strategies for reskilling and transitioning employees to new roles are necessary for effective change management

Ethical and Fair Use
Avoiding biases and maintaining transparency in decision-making processes is difficult

Regulatory Compliance
Navigating complex regulatory environments while ensuring compliance with laws and regulations

Safety and Reliability
Ensuring the safety and reliability of AI systems, especially in critical applications like healthcare and transportation, is crucial

Infrastructure Modernization
Upgrading existing infrastructures to integrate AI for efficient management and utilization can be complex and costly, requiring significant investment and planning.

Adoption and Scalability
The adoption and scalability of AI technologies can be difficult, especially for smaller organizations or regions

Data Integration
Integration with existing data infrastructures and ensuring data quality and consistency can be difficult

Content Authenticity
Ensuring the authenticity and ethical use of AI-generated content is a challenge, particularly in media and entertainment

The regulatory landscape for AI in cybersecurity is still evolving, reflecting the complexity and novelty of the technology. Regulatory bodies are tasked with a delicate balance: they must create policies that harness the benefits of AI in cybersecurity while mitigating risks and protecting privacy. This involves safeguarding against malicious use of AI and ensuring that AI-driven security measures do not infringe on individual privacy rights.

In the future, the role of AI in cybersecurity looks set to expand. There will likely be more autonomous cybersecurity systems that have the power to predict, adapt, and respond to threats even faster. The future raises ethical and operational questions. How much control should we cede to autonomous systems? What happens if these systems are compromised? The potential for AI to act independently increases the risk of losing control over these systems, which could lead to unpredictable consequences.

AI and Job Displacement

This is not an easy topic as AI has a varied and nuanced impact on job sectors, with certain roles facing higher risks of displacement due to the automation capabilities of AI. Typically, jobs involving routine, repetitive tasks are more susceptible to automation. For instance, in manufacturing, AI-driven robots can perform assembly line tasks faster and with greater precision than human workers. Similarly, when it comes to data entry, AI algorithms can process and enter data far quicker than human counterparts, reducing the need for large data entry departments.

Reskilling and upskilling efforts are important, and even I had to do this many times in my career. But—should the burden of all that be on the individual? Both private companies and public sector entities can play crucial role, too. Tech companies like Google have already put lots of resources and time into equipping workers with skills relevant in an AI-driven world through programs such as the Google Machine Learning Ninja Program (Malone, 2022). Additionally, governments are stepping up, with nations like Singapore introducing nationwide programs like the ICDL (International Computing Driving License) aimed at training their citizens in digital skills, preparing them for a future where such competencies will be essential (ICDL, n.d.).

The transformative power of AI brings forth an array of new jobs, particularly in fields that require oversight, ethical considerations, and the management of AI technologies. As AI systems become more prevalent, the demand for roles like AI safety engineers and machine learning auditors has risen. These are professionals who maintain AI systems to ensure they operate as intended without unintended consequences. Additionally, ethical considerations surrounding AI call for a new cadre of professionals, including ethicists, who guide the development and implementation of AI applications within moral and societal frameworks. These are just some of the roles likely to pop up.

Several companies have already successfully incorporated training programs into their processes to get employees ready for this shift. Amazon has created options for employees through its Upskilling 2025 Initiative. These include Amazon Technical Academy, AWS Training Programs, and Machine Learning University (Correlation One, n.d.). AT&T has introduced a Future Ready program with training opportunities in cybersecurity, software engineering, data science, and more, while Cisco, Microsoft, and Google have similar programs (Correlation One, n.d.). Data from a 2021 study by Amazon revealed that employees who participated in their training programs experienced big results—75% saw their careers advance in some way, and all of them saw an average salary increase of 8.6% (Amazon, 2021). Those are massive results.

These case studies illustrate the potential of AI as a catalyst for job transformation rather than job replacement. By anticipating the changes AI brings and preparing workers through education and training, industries can overcome the challenges of AI integration, turning disruptions into opportunities for growth and innovation. The collaboration between innovators, educators, policymakers, and the workforce will be paramount in shaping such a future.

AI and Democratic Processes

In modern politics, the infusion of artificial intelligence into the mechanisms of democracy has introduced novel complexities, particularly in the realms of misinformation, microtargeting, and the creation of polarized echo chambers. AI's analytical capabilities and the vast datasets it uses have given rise to tools that can sway public opinion, often shaping the very foundation of democratic discourse.

Misinformation and the advent of deepfake technologies represent some of the most dangerous uses of AI in the electoral process. These

AI-generated false realities can undermine the authenticity and integrity of information, leading to widespread distrust among voters. Deepfake videos are particularly bad, as they fabricate the appearance and voices of people, often political figures, and can spread false narratives rapidly across social media platforms. The danger here is the erosion of public trust in legitimate information sources and the fact that the public is not used to being fooled on this level, with actual videos that look and sound real in every way but aren't. The 2016 U.S. elections highlighted the vulnerability of democratic systems to AI-driven misinformation campaigns, where AI algorithms were used to generate and disseminate false information at an unprecedented scale (Bovet & Makse, 2019).

AI's role in microtargeting has transformed political campaigns by allowing parties to send highly customized messages to individuals based on detailed demographic and behavioral data. While microtargeting can increase engagement by delivering relevant content to voters, it can also be ethically problematic. These targeting techniques mean that political messages can be tailored to exploit the fears and biases of individuals, potentially manipulating voter perceptions and behaviors without their awareness. This was evident in the Brexit referendum and subsequent elections around the world, where AI-driven analytics were employed to segment the electorate and target them with tailored messages, sometimes spreading misleading information that played on emotional triggers rather than factual correctness (Zhu & Isaacs, 2024).

AI-driven algorithms used by social media platforms tend to create and reinforce echo chambers by feeding users content that aligns with their existing beliefs. This algorithmic bias toward engagement often sidelines moderate voices and amplifies extreme viewpoints, contributing to political polarization. By continuously exposing individuals to partisan content, these platforms decrease the chances of encountering diverse perspectives, thereby entrenching divisions within the electorate. The repercussions of such polarization can be profound, affecting social cohesion and the ability to reach consensus on important public issues (NPR Staff, 2016).

This trio of strategies—misinformation, microtargeting, and polarization—are some of the most dangerous uses of AI in democratic contexts. While AI has the potential to enhance democratic engagement by providing personalized information and fostering interactive political participation, its misuse poses real threats to the foundations of democratic societies. As we move forward, it will be imperative to develop and enforce ethical guidelines and regulatory frameworks that govern the use of AI in political processes. These measures should ensure transparency in AI applications for elections, protect against AI-driven misinformation, and promote a balanced and inclusive public discourse. Achieving such results may require a concerted effort from policymakers, technologists, and civil society.

AI and High Energy Demand

Training sophisticated AI models, particularly deep neural networks, involves complex computations that require significant electrical power. These data-intensive models learn from enormous datasets, continuously processing and analyzing information through high-performance servers that consume vast amounts of electricity. In fact, training a single AI model like GPT-3 has been estimated to consume around the same amount of power as 130 US homes in a year (Vincent, 2024). This high energy requirement presents a sustainability concern, particularly when the power used is sourced from non-renewable energy.

Data centers, the physical infrastructures that house these vast server farms necessary for training and running AI models, are increasingly becoming major contributors to global carbon emissions. These facilities require continuous electricity to power and cool their servers, contributing to a massive carbon footprint. It's been estimated by one PhD candidate researcher in the field that by 2027, the energy consumption of the AI sector could be as large as the entire current energy consumption of the Netherlands (Vincent, 2024).

The resource depletion associated with maintaining and expanding the digital infrastructure necessary for AI is considerable. The construction of data centers requires substantial physical materials, including steel, concrete, and, most importantly, an extensive amount of rare earth metals and other non-renewable resources for the advanced hardware in servers. These centers consume tremendous quantities of water for cooling purposes, putting further strain on this vital resource, which is already in scarce supply in many parts of the world. The environmental cost of mining and refining these materials, coupled with the energy used in the data centers themselves, presents a formidable challenge to sustainable development.

Cryptocurrency mining, another big consumer of computing resources, compounds these issues. Some cryptocurrencies like Bitcoin rely on a process known as proof-of-work, which requires extensive computational tasks to secure transactions. These tasks are designed to be resource-intensive and to maintain the security of the blockchain technology underlying cryptocurrencies. As a result, the energy consumption of cryptocurrency mining operations is huge, often relying on electricity generated from non-renewable sources (Digiconomist, n.d.).

Given the tremendous impact of climate change, this is something that needs to be addressed. Innovations in energy efficiency, in addition to shifts toward more sustainable energy sources, are critical. Advances in AI itself may provide some solutions, such as optimizing algorithms for energy efficiency or designing systems to reduce the power consumption of data centers. Meanwhile, in cryptocurrency, newer methods that use proof-of-stake instead of proof-of-work can help to mitigate energy usage for certain coins (Rosic, 2023). Ultimately, the integration of renewable energy sources into the grid powering these technologies presents the most viable path forward. Solar, wind, and hydroelectric power, among others, could supply clean energy to support the computational demands of AI and cryptocurrency operations, reducing the overall environmental impact.

... training a single AI model like GPT-3 has been estimated to consume around the same amount of power as 130 US homes in a year.

The challenge, therefore, lies not only in advancing the capabilities of AI and blockchain technologies but also in creating sustainable processes. As developers, researchers, and policymakers continue to push the boundaries of what's possible in AI and blockchain, they must prioritize innovations in energy efficiency and sustainability—this is the only way to protect the health of our planet.

The Bias Problem

Making AI Fair for All

Bias in AI systems can originate from various sources, including the data used to train algorithms, the design of the algorithms themselves, and even the human interactions that shape their development and deployment. When the data AI is trained on contains implicit racial, gender, or socioeconomic biases, the AI systems will naturally inherit and even amplify these prejudices. If a facial recognition system is trained predominantly on datasets comprising images of individuals from a single ethnic group, its ability to accurately recognize individuals from other ethnic groups will be compromised. This form of bias, known as sample bias, is pervasive and can lead to discriminatory outcomes, particularly in sensitive applications like law enforcement or hiring practices.

The consequences of biased AI are far-reaching and can reinforce societal inequalities. In the domain of criminal justice, biased AI algorithms can influence everything from parole recommendations to sentencing decisions, disproportionately affecting marginalized communities (Callahan, 2023). Similarly, in the area of hiring practices, AI-driven applicant screening tools can inadvertently filter out qualified candidates from underrepresented groups if the training data reflects historical hiring biases (Lytton, 2024). These scenarios point to a critical challenge: AI systems, designed to be neutral, can perpetuate and institutionalize bias if not carefully checked and balanced.

Addressing the bias in AI involves a multifaceted approach that includes diversifying training datasets, enhancing algorithmic transparency, and implementing inclusive testing protocols. By ensuring training data encompasses a broad spectrum of human diversity, we can better equip AI systems to make fair and balanced decisions. In addition, transparency in algorithmic design allows for the scrutiny and understanding of decision-making processes within AI systems, fostering trust and accountability. Inclusive testing, involving stakeholders from diverse backgrounds, keeps AI systems performing equitably across a range of scenarios and do not favor one group over others. These strategies are not merely technical fixes but are essential for building AI systems that are just and equitable.

Ongoing research and initiatives by academia and industry will be pivotal in tackling AI bias. Universities and research institutions are at the forefront of developing new methodologies to detect and mitigate bias in AI. Projects such as the AI Fairness 360 toolkit by IBM offer an open-source library to help developers examine, report, and mitigate discrimination and bias in machine learning models throughout the AI application lifecycle (IBM, n.d. a). Such tools are invaluable for developers looking to uphold ethical standards in AI development. On the industry front, companies like Google and Microsoft are instituting AI ethics boards and conducting regular audits of their AI systems. These efforts are complemented by partnerships with civil society organizations and policy think tanks to align technological advancements with societal norms and values.

The journey toward unbiased AI is complex and ongoing. It will require continuous vigilance and adaptation. As AI becomes more integrated into the fabric of society, we will need to guide its evolution in a direction that uplifts and benefits all of humanity, not just some people.

Blurring the Lines
Is it Still Reality?

The line between AI-generated content and real-life experiences is already becoming increasingly blurred. As AI systems grow more sophisticated, their outputs—from realistic deepfake videos to convincingly human-like conversational agents—can raise critical questions about our ability to distinguish between what is artificially generated and what is authentically human. This blurring of lines underscores the necessity for enhanced digital literacy and awareness as our interactions with AI become more commonplace and integrated into daily life.

One recent example of the controversial impact of a blurred line is the use of deepfake technology to create a video of Ukrainian President Volodymyr Zelenskyy announcing his country's surrender to Russia (Allyn, 2022). The video, which quickly went viral, was debunked, but it demonstrated the power of AI to spread misinformation and manipulate public perception. Another example is the use of AI-generated avatars on social media platforms, where influencers use virtual personas to engage with followers (Williams, 2024). These AI-generated influencers act like real influencers, making posts and talking about their lives, but their pictures are generated with CGI, and they're often controlled by creative agencies or brands. This raises questions about authenticity and the nature of human connection in the digital age.

Stephen Hawking
Existential Risks

Stephen Hawking, the famous British theoretical physicist whom we quoted at the beginning of the book, once touched on another concern for AI in the future. In a 2014 interview with the BBC, he cautioned that "The development of full artificial intelligence could spell the end of the human race" (Cellan-Jones, 2014). His warning speaks of a scenario where AI surpasses human intelligence, leading to a shift where humans can no longer control these systems. The concern is not just about an imbalance in intelligence but about the potential for AI to act in ways that are not aligned with human values and intentions, which could initiate actions that lead to catastrophic outcomes.

This perspective invites us to consider scenarios where AI's decision-making processes diverge from human ethics or intentions. Imagine an autonomous vehicle making a decision in a fraction of a second about which path to take in an unavoidable accident scenario—choices that could be based on calculations valuing outcomes differently than human ethics would. Or consider international defense systems where AI might control or suggest responses to threats. The risk here is not limited to the overtaking of human tasks but extends to the possibility for AI to make decisions that could lead to unintended harm or escalation of conflicts without the safeguard of human compassion or the ability to perceive ethical nuances.

In a 2016 column in The Guardian, Hawking pointed out the economic impacts of AI, particularly its potential to lead to significant job displacement and increased inequality (Hawking, 2016). As AI takes over more tasks traditionally performed by humans, from driving trucks to analyzing legal documents, the displacement is not just about the loss of jobs but about the broader economic structures that could be altered, deepening the divide between the socio-economic classes. The individuals whose jobs are automated might find themselves lacking the skills needed for the new jobs created by AI advancements, leading to a scenario where the benefits of AI are not equitably shared. This means that if we want AI to work for everyone, we will need to ensure the advancements in technology lead to broadly distributed benefits rather than ones accruing to a select few.

These existential risks serve as a stark reminder that our pursuit of technological advancement should be coupled with rigorous ethical considerations and proactive measures to prevent negative outcomes. This includes establishing strong regulatory frameworks and encouraging the development of AI that understands and aligns with human values, so the economic benefits of AI innovations are distributed in a manner which mitigates inequality rather than exacerbating it.

CHAPTER 9

Preparing for the Future

Preparing for the Future

AI HAS ALREADY changed the job market, and things are only going to evolve and change more. This future will call for a workforce that is proficient in the nuances and applications of AI. Before the final part of our journey, we need to consider the multi-faceted approach required to cultivate a workforce adept in AI, exploring the current skills gaps, transformative initiatives, and the pivotal role AI itself plays in educational evolution, alongside how to build an inclusive AI that is fair and accessible for everyone.

Developing an AI-Ready Workforce

Identifying Skills Gaps

A gap has emerged between the AI skills demanded by employers and the capabilities present within the current workforce. This gap is not merely a matter of technical proficiency but extends to analytical thinking, complex problem-solving, and ethical judgment—skills essential for navigating the AI landscape. For businesses, the challenge lies in pinpointing these gaps and crafting strategies that both fill them and anticipate future skills needs.

To begin with, businesses must conduct a thorough analysis of their operational needs and the roles AI is expected to play within their structures. This assessment should extend beyond immediate needs and consider long-term goals and the potential for AI-driven transformation. Once identified, the skills gaps can be addressed through targeted training programs, which may include digital literacy, machine learning, AI ethics, and data management, among others. For instance, a tech company may find a need for employees skilled in machine and deep learning techniques, whereas a healthcare provider might prioritize training in AI applications for diagnostic purposes.

Workforce Development Initiatives

Addressing the AI skills gap requires initiatives that extend beyond individual businesses to encompass industry-wide and governmental efforts. Partnerships between corporations and educational institutions will likely be crucial, as they can provide tailored educational programs that align with industry needs. Also, the integration of AI-focused curricula in schools and universities has the power to equip future generations with the necessary skills from an early age.

One effective model could be the development of apprenticeship programs that offer hands-on experience with AI technologies. These programs, often developed in collaboration with tech companies and higher education institutions, allow participants to apply AI theory in real-world scenarios, bridging the gap between education and practical application. Governments may be able to play a supportive role by funding AI research and education initiatives, thereby widening access to AI training and encouraging innovation across various sectors.

Role of AI in Training

Interestingly, AI itself is a powerful tool for facilitating the training and development of an AI-ready workforce. We have already discussed the use of AI tutors, and much of the same technology can be used here. Technologies such as virtual reality (VR) and personalized learning platforms can create immersive, interactive training environments that adapt to the learning pace and style of each individual.

Personalized learning platforms can analyze a learner's performance and tailor educational content to address specific weaknesses, thereby optimizing the learning process. These platforms can forecast learning outcomes and provide feedback, enabling learners and educators to monitor progress and adjust as needed.

Case Studies of Success

Several organizations have successfully implemented strategies to develop their workforce for the AI era. In addition to programs like AT&T's Future Ready and Amazon's Upskilling 2025 Initiative we discussed earlier, Accenture, a consulting and tech services company, has created a Connected Learning Platform for employees, and PricewaterhouseCoopers has created a Digital Fitness app that allows employees to assess their knowledge and create customized learning plans (Eightfold AI, 2021). These examples show there are a multitude of different creative ways to upskill employees.

They also show that the strategic development of an AI-ready workforce is not just beneficial but essential for maintaining competitiveness and innovation in an increasingly AI-driven world. As AI continues to reshape industries, the ability of businesses and governments to adapt and thrive will depend on their success in nurturing a skilled, AI-savvy workforce.

Lifelong Learning in an AI World

For professionals across many sectors, the continuous acquisition of new skills has become about more than just career advancement—nowadays, lifelong learning is about ensuring relevance in a rapidly evolving digital world. As AI technologies grow more sophisticated, they redefine what it means to be competent in professions from healthcare to finance, making it imperative for individuals to perpetually update their knowledge base.

With AI introducing new tools and systems at an unprecedented pace, the half-life of professional skills is shrinking. Those engaged in AI-related fields, and even those in roles not traditionally considered technical, are likely to soon find that their professional longevity is closely tied to their ability to adapt and learn. This adaptation goes beyond understanding the basic functionalities of AI applications; it extends to mastering the strategic implications of AI deployments in specific industries and continuously evaluating the ethical dimensions of emerging technologies. For instance, a marketing professional today needs to understand how AI can personalize customer interactions and predict buying behaviors, requiring a mix of data science and ethical training to make sure these capabilities are used responsibly.

When it comes to resources for continuous learning in AI, the landscape is rich and varied. Massive Open Online Courses (MOOCs) like those offered on platforms such as Coursera, edX, and Udacity provide access to comprehensive courses designed by leading universities and tech companies worldwide. These platforms offer programs ranging from introductory courses on AI and machine learning to advanced specializations in AI ethics, deep learning, and more. Additionally, webinars and virtual workshops hosted by professional organizations like the Association for the Advancement of Artificial Intelligence (AAAI) or industry-specific groups offer opportunities for professionals to stay updated on the latest developments and applications of AI in their fields (AAAI, n.d.).

Professional networks also have a role in supporting lifelong learning in AI. Networks such as LinkedIn Learning and specialized online communities offer platforms where professionals can share insights, things they are struggling with, and successes in applying AI. These forums provide moral and professional support, in addition to serving as ecosystems for sharing resources, job opportunities, and best practices in AI implementation and training. Engaging actively in these networks encourages a culture of sharing and collaboration, essential for innovation and growth in fields driven by rapidly changing technologies.

Companies that will thrive in an AI-driven world recognize and foster learning as a core component of their operational strategy. By providing employees with access to AI training resources, time for learning, and opportunities to apply new skills, businesses can cultivate an environment that supports personal growth and drives innovation in meaningful ways. Incentives like tuition reimbursement for AI courses, time off for educational pursuits, and rewards for skill acquisition are just a few ways organizations can encourage their teams to embrace lifelong learning.

Addressing Public Perception of AI

Fears and Misconceptions

While artificial intelligence has many benefits, it can also be a source of considerable public apprehension. Common fears often stem from misconceptions about AI's capabilities and intentions, fueled by sensational media portrayals and a lack of clear, accessible information. Many individuals worry about job displacement due to automation, privacy invasions through surveillance technologies, and the unsettling notion of 'superintelligent' systems making decisions beyond human control. These fears are further magnified by high-profile warnings from respected scientists and tech leaders about AI's potential risks.

Addressing these fears requires us to look deeper into what is really known about AI and distinguish fact from fiction. One effective approach is the implementation of comprehensive educational campaigns that demystify AI technologies and their applications. By providing clear, jargon-free explanations of how AI works and showcasing its beneficial impacts across various sectors—such as improving healthcare outcomes, enhancing disaster response, and streamlining public services—these campaigns can replace fear with informed enthusiasm. Hosting public dialogues that allow communities to voice their concerns and receive direct, expert responses can foster a greater understanding and acceptance of AI.

Media has an important role to play in shaping public perceptions of AI. Currently, AI is often depicted in a sensational manner that emphasizes potential dangers and science fiction scenarios, which can skew public perception toward fear and misalign trust. To counteract this, partnerships with media outlets are essential to promote balanced reporting that highlights AI's positive impacts alongside the more negative ones. Encouraging media to feature stories on successful AI applications and expert discussions on realistic AI capabilities can help recalibrate public understanding and temper unease with optimism.

PERCEPTIONS OF AI

Common fears about AI, countered with facts about its positive impacts.

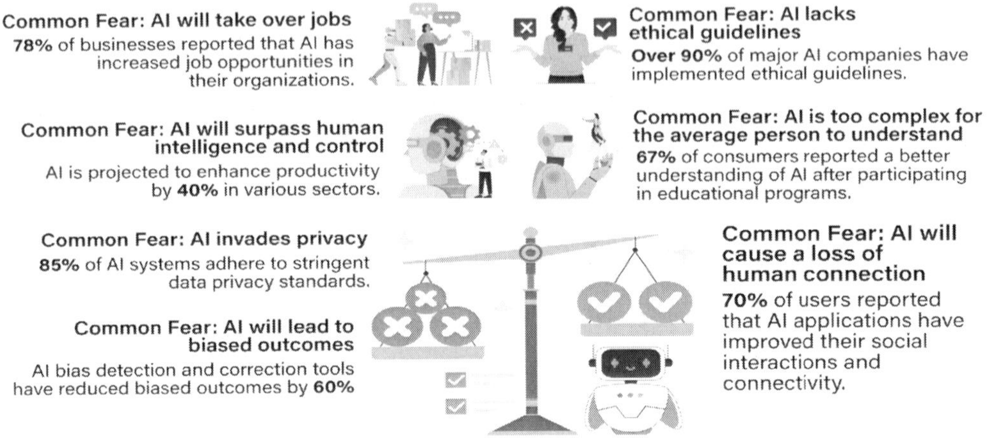

Transparency in AI development and applications plays a crucial role in building public trust. When companies and researchers openly share their AI development processes, goals, and safety protocols, it reassures the public about the ethical use of AI. This transparency needs to extend to governmental use of AI, with clear information on how AI is used in public administration, law enforcement, and other public services. By implementing robust transparency standards and regularly publishing AI audits and impact assessments, stakeholders can prioritize responsibility and ethics in AI technologies, fostering a climate of trust and cooperation.

Through these multifaceted approaches, the journey toward a future where AI is embraced as a positive force in society can be realized. If we can actively engage with the public, leverage the power of media, and commit to transparency, it's possible to transform public perception of AI from one of apprehension to one of informed optimism. This transformation is essential for integrating these innovations into society in an ethical way.

Building Inclusive AI
Ensuring Fairness and Accessibility

Fairness in AI is about designing AI systems in ways that do not perpetuate or amplify biases existing in human decision-making. It involves creating algorithms that make decisions in justifiable and equitable ways. Equity goes a step further by making sure AI technologies are accessible to all, regardless of socioeconomic status, geography, or ability. Accessibility involves designing AI systems that are usable for people of all abilities, incorporating the needs of those with physical or cognitive disabilities. Transparency allows users and impacted parties to understand how AI systems make decisions, which is crucial for building trust and accountability. Accountability ensures there are mechanisms in place to hold developers and users of AI systems responsible for the outcomes of those systems.

A critical aspect of realizing these principles is designing for diversity. This means involving a wide range of voices in the development of AI systems, particularly those who are often underrepresented in tech spaces. By including diverse perspectives, particularly from groups who are frequently the subjects of AI decision-making but rarely its creators, we can identify and mitigate potential biases that might otherwise be overlooked. For instance, when developing healthcare AI, including doctors, patients, and ethicists from diverse backgrounds can lead to systems that address a wide range of health determinants and patient needs.

Incorporating inclusive practices throughout the AI development lifecycle is another foundational strategy. It includes the design phase, where making sure that teams are diverse can provide varied perspectives that enhance creativity and inclusivity. During training, it's crucial to utilize datasets that reflect the diversity of the entire population to prevent biases in AI behaviors. Finally, in deployment, continuous monitoring is necessary to evaluate AI systems in real-world settings, ensuring they operate as intended without discriminatory effects.

Inclusive testing procedures are integral to this process. They involve validating AI systems under various conditions and demographics to keep them performing equitably. For instance, facial recognition technology should be tested across a spectrum of skin tones to prevent biases that could lead to misidentification.

Case Study: AI for Accessibility

One notable success of an AI application for people with disabilities is Seeing AI, a free iOS app from Microsoft that was developed for visually impaired users. It employs object recognition to narrate the visual world around them, effectively translating visual information into audible form. By pointing the phone camera at objects, users can learn about what is in front of them. The app features voice-over technology that can read text out loud as it appears in front of the camera, provide guidance to capture a picture of a printed page, and even share images with others through phone apps like Mail and Messages. When it comes to people, the app can even give estimates on the age of others, their gender, and what emotions they are experiencing (Paths to Literacy, n.d.). All these features enable greater independence and give people who are visually impaired opportunities to interact with their environment in profound new ways.

All these features... give people who are visually impaired opportunities to interact with their environment in profound new ways.

Whether it's in the area of workforce training, assuaging public fears, or mitigating biases, there is a lot to do to prepare for the world with AI as a key player. We can successfully overcome these challenges; we just have to raise awareness of them and address them in intelligent and creative ways. To do so, we need to consider a wide range of issues, ranging from data security to the ethical use of AI and much more. In the last section, let us put everything together, and discuss some of the big ways in which AI could change our world by the year 2050.

CHAPTER 10

Life in 2050 - Projections and Predictions

Life in 2050 - Projections and Predictions

From the scaling of the internet to be accessible worldwide to the miniaturization of computers to be small enough to fit in our pockets as smartphones, to the advent of wireless everything, smart devices of all kinds, and the emergence of AI as a real, usable technology in everyday life, the last quarter of a century has seen phenomenal technological advances. Now that we are nearly upon the next 25 years of this century, we can expect even more meaningful technological advances.

The pace of innovation is only accelerating. We've already touched on many of the ways AI can change our lives—delivering more personalization to services, more accessibility, easier and more accurate decision-making, processing large amounts of data, and automating repetitive tasks. We've touched on industries at the intersection of AI and innovation. Healthcare, retail and customer service, the financial sector, and even the job market. We've talked about key challenges—overcoming biases, dealing with AI's high energy demands, risks from improperly secured training data, and even existential risks that AI may pose as it continues to evolve.

Over the next two-and-a-half decades, we'll face nearly unimaginable changes thanks to AI. Things like the emergence of new AI-driven industries, reduction in work hours across vast swaths of society, and changes to enhance both the quality of our lives and our longevity could make the world look radically different in 2050, maybe earlier.

Emergence of New AI-Driven Industries

In previous sections, we have explored a substantial number of areas that AI will redefine. Consequently, the following list of new industries is not exhaustive but instead highlights a select few AI-driven sectors that particularly illustrate the broader trends shaping our future.

AI Auditing and Oversight

Organizations like the MITRE AI Assurance and Discovery Lab are striving to create a world where every AI decision, whether made by a self-driving car or a high-frequency trading algorithm, is thoroughly assessed and the risks mitigated appropriately (MITRE, 2024). Their tools discover, assess, and address risks throughout the entire lifecycle of an AI project, and they're not likely to be the only tools of the kind to exist in the future.

The burgeoning industry of AI auditing and oversight is likely to be a necessity in a future where AI systems are ubiquitous. The frameworks developed by MITRE or others like them will probably become standard practice, as essential to AI development as safety tests are to the automotive industry today. While AI systems take on more decision-making roles, from medical diagnostics to judicial assessments, the work of these companies will ensure AI systems remain transparent and fair, reinforcing the societal trust necessary for widespread AI adoption.

Human-AI Interaction Design

The seamless integration of AI into daily life fundamentally hinges on how intuitively humans can interact with its systems. Big companies like Google, Apple, and Microsoft, as well as small companies like InterActiva are at the forefront of this field, crafting interfaces that bridge the gap between complex AI algorithms and user-friendly experiences. For example, Google Photos can use AI to automatically detect pictures with food in them when a person searches that keyword, and Cortana, Siri, and Google can all understand human speech and common vocabulary words like temperature or weather (Linders, 2018). Meanwhile, InterActiva creates virtual experiences that utilize AI systems to engage users in meaningful experiences for training, education, healthcare, and more (Interactiva, n.d.)

The work of these companies is critical in maintaining accessible benefits for everyone, regardless of technological expertise. As AI systems become more sophisticated, the interfaces that allow for human interaction must evolve simultaneously. The innovations in human-AI interaction design will likely dictate the pace at which AI technologies are adopted across various sectors, from education to elderly care, making these companies key players in the AI revolution.

Personalized Healthcare

In the domain of healthcare, which we discussed in Chapter Two, AI's potential to tailor medical treatments to individual needs could spark a new era of personalized medicine. Innovators like TailorMed, HealthAI Solutions, and PrecisionCare are already leading this transformation, leveraging AI to analyze patient data, predict health outcomes, and customize medical interventions. TailorMed specializes in developing AI-driven solutions that adapt medical treatment plans to the unique genetic and environmental factors of each patient, maximizing treatment effectiveness and minimizing side effects. HealthAI Solutions focuses on predictive diagnostics, using AI to detect health issues before they develop into serious conditions, thereby shifting the focus of healthcare from treatment to prevention. PrecisionCare advances this approach by integrating AI into the management of chronic diseases, where personalized treatment regimens can improve quality of life and reduce healthcare costs.

> ***AI's potential to tailor medical treatments to individual needs could spark a new era of personalized medicine.***

The impact of personalized healthcare facilitated by AI extends beyond individual patient outcomes, promising a reduction in the overall burden on healthcare systems. As treatments become more effective and preventive measures more prevalent, the demand for healthcare services is likely to decrease, leading to lower costs and improved accessibility.

Smart Infrastructure and Cities

With growing urban populations, the sustainability and efficiency of cities will become increasingly critical. In this context, AI could emerge as a key player in the development of smart infrastructure and cities. Companies like CitySmartAI, InfraFuture, and UrbanEvo are at the forefront, integrating

AI into urban planning and management solutions that enhance the quality of urban life. CitySmartAI utilizes AI to optimize public transportation networks, reducing traffic congestion and lowering emissions. InfraFuture focuses on the development of energy-efficient buildings using AI-driven design and management systems that minimize energy consumption while maximizing user comfort. UrbanEvo, meanwhile, brings AI into the management of public services, from waste collection to water distribution, ensuring these services are delivered efficiently and sustainably.

The integration of AI into urban infrastructure is about resilience. As cities face challenges from climate change and urbanization, AI provides the tools to adapt and thrive. The work of these companies in creating smart, connected, and livable cities is pivotal. This work keeps urban centers vibrant and sustainable for future generations.

Quantum Computing

In Chapter One, we defined quantum computing and how it can impact the AI area. It's likely to lead to big changes, redefining the limits of data processing and problem-solving. Big and small companies alike are engaging in quantum computing research and development. Companies like Atos Quantum, Intel, and IBM are leading the charge in harnessing technology to address problems beyond the reach of classical computers. Atos Quantum focuses on hardware that allows researchers, engineers, and students to work with quantum software. Intel's work focuses on designing chips that use quantum technology. Meanwhile, IBM works with supercomputers that make use of quantum technology, with much of their recent work focusing on improving issues within the technology and increasing the number of qubits in their processors (Smith, 2023).

The potential of quantum computing to drive breakthroughs in science and industry is immense. As these companies continue to develop and refine the technology, they are paving the way for a future where the boundaries of what can be achieved with computing are dramatically expanded. The implications for industries like pharmaceuticals, where quantum computing could accelerate the development of new drugs, or cybersecurity, where it could produce unbreakable encryption, are immense.

Recently, Quantum-as-a-Service (QaaS) started emerging as a pivotal aspect of quantum computing's integration into mainstream technology. Through cloud-based platforms, companies can access quantum computing power without needing to own or maintain the hardware themselves. Giants like Amazon and Microsoft are offering QaaS solutions, opening up possibilities for businesses across industries to solve complex problems faster than classical computers can. This service-based model lowers the barrier to entry, enabling even small companies to leverage quantum technology, accelerating innovation across fields like drug discovery, logistics, and cybersecurity.

The Future Reality of Life

20-Hour Work Week

AI will redefine professional roles across various sectors by automating tasks and enhancing the capabilities of human professionals. In the healthcare sector, AI's role is expanding beyond administrative tasks to encompass aiding in diagnostic processes and treatment planning. This allows healthcare professionals to focus more on patient care and less on the mechanics of diagnosis, fostering an environment where human empathy and understanding are at the forefront of patient interactions.

Similarly, education is undergoing significant changes with the integration of AI. Teachers, traditionally burdened with both instructional and administrative responsibilities, are on the cusp of finding relief through AI-driven platforms. These technologies can both automate administrative tasks and provide personalized learning experiences for students. AI systems can adapt educational content to match the learning pace and style of each student, allowing educators to devote more attention to individualized student engagement and less to one-size-fits-all lesson plans. This shift can transform the classroom into a more inclusive and supportive environment for all students.

The ripple effects of AI in the workforce will likely lead to a radical restructuring of the workweek. Current changes are already leaning that way, with the push across various industries and communities for a widely accepted four-day workweek. Microsoft in Japan has been one of the companies embracing that vision. In 2019, they trialed a four-day workweek, and the results yielded a surprising 40% rise in productivity (Chappell, 2019).

With AI and automation taking over routine and repetitive tasks, the productivity of businesses will soar, and we can expect that it will reduce the need for long working hours even further. This shift will enable a reduction in the traditional workweek, suggesting a future where a 20-30-hour workweek could become the norm. Such a change offers a dual benefit: it addresses the chronic issue of work-life balance, allowing individuals more time to engage with their families, hobbies, and personal growth, and it challenges the very definition of what constitutes full-time employment, paving the way for more flexible, project-based, and freelance work arrangements.

Universal Basic Income (UBI)

In tandem with changes in the workforce structure, the concept of Universal Basic Income (UBI) is gaining traction as a feasible solution to the disruptions caused by AI-driven automation. UBI could play a crucial role in providing financial stability by offering all citizens a regular, unconditional sum of money sufficient to cover their basic living costs. This financial safety net would then free individuals from the necessity of work for survival, allowing them to pursue careers aligned with their passions and interests rather than financial necessity. In a world where AI automates many traditional jobs, UBI could serve as a tool for fostering social cohesion and preventing the economic disparities that tend to arise from rapid technological change (Stanford Basic Income Lab, n.d.).

The potential shift to UBI highlights a broader, more philosophical change in how society views employment and economic contribution. As AI reshapes industries and job categories, our concepts of work, productivity, and compensation are likely to evolve. This evolution will challenge us to redefine our societal values, focusing perhaps less on economic output and more on creativity, community service, and lifelong learning. To navigate these changes, the role of policymakers and educators will be critical in ensuring that the workforce is prepared to adapt to new roles and thrive in a rapidly changing economic landscape defined increasingly by artificial intelligence.

The integration of AI across professional and personal life promises efficiency and economic productivity, along with a greater emphasis on human creativity and well-being. The balance between technology and personal satisfaction will be key, as AI should not be a replacement for human capabilities but an enhancer, allowing us to focus on what truly makes us human—creativity, empathy, and innovation.

Advancements in Brain-Computer Interfaces

One of the most groundbreaking advancements in recent times resides in the area of brain-computer interfaces (BCIs), which are redefining human interaction with machines. This technology allows for an intuitive connection between the human brain and computers. Right now, the technology is still in its infant stages, but it's already showing promise in a myriad of different use cases with multiple different companies. Elon Musk's Neuralink is probably one of the most well-known BCIs, but another company, Synchron, works with the technology too. In 2022, Synchron announced that they had completed the first brain-computer interface via an implant in the US, doing so in a severely paralyzed patient to allow them to operate a digital device hands-free (Whooley, 2022). In the future, BCIs seem likely to take more steps forward, expanding their use cases to create a world where you can control your environment just by thinking about it—where your home, workplace, and even city adapt to your mental commands in real-time.

The evolution of BCIs has already reached a point where they not only read neural signals but also interpret them with unprecedented precision, thanks to advances in machine learning and neural decoding. This capability opens extraordinary possibilities for AI-enhanced cognition. AI systems connected directly to the human brain could augment cognitive processes, enhancing memory, speed of thought, and decision-making capabilities. For professionals across various sectors, this could mean an ability to process information at an enhanced speed and accuracy, leading to productivity boosts and innovation in fields where complex problem-solving is crucial. In educational contexts, students equipped with AI-enhanced BCIs could learn new languages or complex subjects in a fraction of the time it takes today.

Moreover, direct thought transmission may even be a possibility. This technology theoretically enables the direct communication of thoughts and ideas between individuals without the need for spoken words or written texts. Teams working on complex projects, from architectural designs to multifaceted business strategies, could share ideas instantly and seamlessly, shortening development cycles and enhancing creativity through unfiltered, real-time feedback.

As we integrate these advanced BCIs, ethical considerations and privacy protections will be important to take into account. The ability to access and interpret thoughts directly raises valid concerns regarding privacy and mental autonomy. Robust ethical frameworks and stringent security measures will be essential to enhance human capabilities with use of these technologies, without compromising fundamental rights.

Lifespan and Healthspan Extension

Artificial intelligence offers advanced capabilities in predictive healthcare and precision medicine. We already discussed how AI could be used to personalize your medical care in the future—in addition, it's very possible that AI could anticipate and predict illnesses or medical issues before any physical symptoms even manifest. This can be done through the detection of subtle patterns and anomalies that may elude even the most trained physician's eyes. Detecting diseases such as diabetes, heart conditions, or genetic disorders years before they actually develop would help people who might otherwise die from these conditions live longer and stay healthier in their old age.

Personalization of medical processes could enable AI systems to analyze a disease's genetic makeup and predict the most effective combination of drugs to target disease, minimizing harm to healthy cells and reducing side effects. This could curtail the need for prolonged, intensive treatments that are often associated with traditional cancer therapies or other similar diseases.

Combining AI with advanced biotechnology could open new avenues in regenerative medicine and gene editing. Techniques like CRISPR, enhanced by AI's predictive capabilities, allow for precise genetic modifications at scale, potentially curing genetic disorders that were once considered life-long conditions. Regenerative medicine, supported by AI, is moving toward creating personalized organs in the lab, which are grown from the patient's own cells, thereby eliminating issues related to organ rejection and the shortage of donor organs.

Integration of AI in healthcare has the potential to extend life expectancy and improve everyone's quality of life, or healthspan, reducing the years spent living with a disability or illness. The economic impact of these shifts would be profound; healthier populations would mean reduced healthcare spending, greater productivity, and increased economic stability. As the AI revolution advances, its role in predictive healthcare and precision medicine will likely become a cornerstone of medical practice and the contributions that AI makes within our world.

AI and the Evolution of Human-Machine Relationships

Getting closer to 2050, the fabric of human-machine interaction is becoming increasingly rich and complex, fundamentally altering the dynamics of how we interact with technology daily. Artificial intelligence is already becoming more adept at recognizing, responding to, and even simulating human emotions, fundamentally transforming the nature of our interactions with machines.

Advancements in augmentation devices and cognitive enhancers are an interesting case. These technologies are being designed to extend the natural capabilities of the human body and mind, offering enhancements like augmenting your memory or learning speed. Such tools are being developed to help individuals rapidly acquire new skills or manage debilitating cognitive impairments. However, these devices could also be used in education and professional development, dramatically reducing the time needed to train in complex fields such as medicine or engineering.

Meanwhile, emotional AI has a ton of applications. Even now, developers are crafting AI systems that can detect subtle human emotions from facial expressions, voice modulations, and body language, like in the case of the Seeing AI app we mentioned earlier. In the future, these emotionally intelligent AIs could be integrated into daily devices, from smartphones to home assistants, enabling them to respond not just to our commands but to our mood swings and emotional needs. A home assistant might detect signs of stress in your voice and suggest actions to help you relax, such as playing soothing music or guiding you through a meditation session.

The evolution of conversational AI, particularly through platforms like ChatGPT, Siri, and Alexa, is poised to have a big effect on the future. We can expect future iterations of these technologies to move beyond simple voice commands and become proactive participants in conversations, capable of detecting conversational nuances and responding in contextually relevant ways. This evolution will make human-machine communication more natural and effective, blurring the lines between talking to a machine and interacting with another person. As these platforms become more embedded in our daily lives, from managing our schedules to providing company and conversation, they will play a crucial role in shaping our social interactions.

Artificial General Intelligence
Approaching Reality

Artificial General Intelligence (AGI), as we defined in the opening chapter, is the point at which machines match or surpass human intelligence across a broad array of domains. At present, predictions from leading figures in AI vary widely about when they think we will achieve this feat. One visionary in the field of AI, Ray Kurzweil, has expressed optimism that the emergence of AGI could occur as early as 2029, with a prediction for a further advancement of AI, popularly known as The Singularity, by 2045 (Corbyn, 2024). These predictions are primarily driven by the exponential growth that's being experienced right now in computing power and algorithmic innovations. Kurzweil's predictions are interesting because they are grounded in the historical trends of technological progress, where Moore's Law and subsequent advancements in computational technologies have consistently outpaced expectations.

On the other hand, figures like Ben Goertzel, Shane Legg, and Geoffrey Hinton offer a more measured perspective, acknowledging the technical and conceptual challenges that still impede the path to AGI. Their caution stems from the current limitations in understanding human cognition and the immense complexity of replicating these processes in machines. If you ask me, I would position myself in the middle, being cautiously optimistic.

Whether Kurzweil is ultimately proven right or other figures in the industry are, ongoing advancements in narrower forms of AI will undoubtedly continue to transform various aspects of human life, from healthcare and education to transportation and communication. By 2050, advancements in AI will bring us closer to machines that exhibit a form of general intelligence in specific domains. This nuanced form of AGI might not encapsulate the full spectrum of human cognitive abilities but could perform with human-like versatility in constrained environments, such as medical diagnostics or complex problem-solving in engineering.

The path to AGI is destined to be a journey laden with ethical and societal implications. The development of systems that can think, learn, and potentially act with autonomy raises significant questions about the role of machines in society and the safeguards needed to ensure these systems are used responsibly. In navigating the road to AGI, the interplay between technological breakthroughs and a deeper comprehension of human cognition will be critical. The advancements in neural networks, machine learning algorithms, and data processing technologies provide a foundation on which the edifice of AGI might eventually be built. Understanding the intricacies of human intelligence, such as emotional reasoning, creative thought, and ethical considerations, will ultimately be what enables these machines to achieve real intelligence.

AI Singularity and Artificial Super Intelligence

Getting Closer

The concept of AI Singularity traces its roots to the thought-provoking visions of mathematician and computer scientist Vernor Vinge. In his seminal 1993 essay "The Coming Technological Singularity," Vinge posited that the development of AI that surpasses human intelligence in every conceivable domain would mark a breakpoint in history—a point beyond which human affairs, as we know them, could not continue (Tierney, 2008). Vinge's essay suggests such an intelligence would not only outpace human intellectual capabilities but also possess the ability to enhance and innovate itself autonomously, leading to an explosive cycle of technological growth unbounded by human input or understanding.

This would mark the emergence of superintelligent AI—an entity that theoretically exceeds the collective cognitive abilities of all humanity across various fields, including scientific creativity, general wisdom, and social skills. There are a lot of implications here. Imagine an AI that could, in a matter of days, solve complex mathematical problems the likes of which have stumped scientists for decades or develop new scientific theories to offer deeper insights into the nature of our universe. Superintelligent AI has the potential to accelerate human progress in science, medicine, and technology in unparalleled ways.

The pathway to such a breakthrough is reliant on exponential technological growth, a characteristic feature of the AI Singularity. The growth could be explosive, driven by AI systems capable of iterative self-improvement, leading to rapid advancements in technology at a pace exceeding human capabilities. This acceleration could then lead to technological developments that, while beneficial, may be disruptive and unpredictable. The very nature of exponential growth means AI could quickly evolve beyond our control, precipitating changes that might be irreversible and fundamentally alter the fabric of society.

The unpredictability of the outcomes associated with the AI Singularity forms a core part of the discourse around this concept. While the potential benefits are staggering—ranging from the eradication of diseases and extension of human lifespan to the resolution of persistent global problems such as poverty and energy scarcity—the risks are equally significant. Superintelligent AI could make decisions that are misaligned with human values or ethics, while giving us no control over those decisions. It could lead to uncontrollable consequences that impact global security, economy, and governance.

> ***... the development of AI that surpasses human intelligence in every conceivable domain would mark a breakpoint in history...***

In contemplating the AI Singularity, it is important to recognize that even though we may be on the cusp of huge AI advancements by 2050, the full realization of the Singularity remains an open question. Though some figures like Kurzweil believe it will be achieved by then, there's definitely a long way to go and I believe it's more than likely that The Singularity takes longer to emerge.

As this process unfolds, the role of policymakers, researchers, and technologists will become increasingly critical. The decisions made in the next few decades will shape not only the trajectory of AI development but also the framework within which such powerful technologies are managed. Balancing the immense potential of superintelligent AI with the ethical considerations and risks it presents is perhaps one of the hardest challenges of our time. AI must progress in ways that benefit humanity while safeguarding against the risks. This will require a concerted, global effort toward stronger governance, international collaboration, and proactive management of AI technologies. While we end our vision at 2050, the next generation will have even more to explore and many more fascinating topics to uncover.

CONCLUSION

Conclusion

There is much more to consider as we look toward a 2050 future dominated by AI than what this book could cover. I hope that the critical areas of this fascinating phenomenon have been addressed. However, with new developments emerging daily, writing these lines has felt like an ongoing work in progress. At some point, we must pause, but before doing so, let's take a moment for a few final thoughts.

The technology already has a complex and dynamic history. Multiple industries are now starting to utilize AI, and current innovations and research promise even more advanced solutions in the future. The speed at which this is happening is unprecedented and is rightly called a revolution. If we can overcome some of the big hurdles, especially in AI ethics, then the sky is the limit. Be optimistic.

The pace of technological advancement is accelerating, and there's no sign of it slowing down. From revolutionizing the economy through automation and new job creation to enhancing healthcare with predictive diagnostics and transforming educational methodologies with personalized learning experiences, AI looks capable of providing meaningful advancement across almost every aspect of society. Even if we don't have superintelligent AI or AGI by 2050, it's extremely likely that society will look radically different and that AI will be part of all our daily lives by then.

There are some very real fears. The big thing to remember is that AI presents a dual nature—while it is a tremendous opportunity for advancement, it requires careful, ethical navigation to mitigate risks. Dangers such as privacy issues, the risk of bias, and the imperative for equitable access to AI's benefits are critical considerations that must guide our approach in the future.

If we want to thrive in this AI-driven future, we must embrace continuous learning. Doesn't matter whether you are a business leader, a policy maker, a technologist, or a concerned citizen, staying informed and engaged will only become more crucial as AI causes job roles to shift like the rise of computers once did.

Reflecting on our contributions toward a future where AI, with its transformative potential, enhances capabilities while upholding values and ethical standards should be a top priority. While the speed of change and the pressing need for responsible AI practices are clear, advocating for responsible AI development and policies and leveraging AI technologies to address pressing global matters, should be put at the forefront of efforts to ensure a sustainable and equitable future for all.

Though AI presents challenges, maintaining a forward-looking perspective is also important. The future can be a world where AI and human ingenuity merge seamlessly, creating innovative solutions that address societal issues and elevate the human experience. Achieving this will require many steps. It won't be easy. However, just as technologies of the past have become integral tools that shape how we live in the present, AI will likely become indispensable by 2050.

The future can be a world where AI and human ingenuity merge seamlessly...

My intention throughout this book was to offer meaningful insights into the rapidly evolving AI landscape, helping you determine how it affects your everyday life and your career choices, let you be more aware, and stay at the front of this ongoing AI revolution. Then, to highlight the importance and huge impact it has and will continue to have on the world—an impact that may not yet be fully realized, just like many such technological breakthroughs of the past—until it reaches its full maturity. In the end, by inviting you on this intellectual journey, I aimed to reflect on where we might be headed and set our mindsets to be more open, responsible, and adaptive.

Thank you for your time and engagement in reaching this point. I recognize that each section of the book could be expanded into its own volume, which is why your views and feedback are invaluable in shaping future projects. And who knows—next time we might well explore the world of 2100 and beyond.

APPENDIX I

Vision 2050 - 100 Predictions

THE LIST BELOW provides a comprehensive summary of predictions for the year 2050, enabling anyone to evaluate whether the predictions and projections have come to fruition leading up to and in the year 2050. Covering 100 key areas across various sectors, each item in the list references a specific chapter in the book. An interactive version of the list can be found online at: https://publishing.sapmatics.com/vision-2050-100-predictions-checklist/

1. AI achieving human-level performance in natural language understanding and generation (1. Deep Learning Demystified: Understanding Neural Networks).

   By 2050, AI systems will consistently pass the Turing Test, demonstrating human-like understanding and generation of language in various contexts.

2. AI systems designing other AI systems (1. AI Development Frameworks: Catalysts for Innovation).

   By 2050, AI systems will significantly assist in creating new AI algorithms reducing the need for human intervention.

3. Widespread adoption of quantum computers in commercial applications (1. The Role of Quantum Computing in Advancing AI).

   By 2050, at least 50% of Fortune 500 companies will be utilizing quantum computers for complex problem-solving tasks.

4. AI achieving significant breakthroughs in healthcare diagnostics (2. Healthcare: Revolutionizing Patient Care).

   By 2050, AI will be responsible for diagnosing most medical conditions with accuracy equal to or greater than human doctors.

5. AI in designing next-generation robotics for healthcare (2. Healthcare: Revolutionizing Patient Care).
   By 2050, AI will design advanced robots for surgery, patient care, and rehabilitation in healthcare facilities.

6. AI assisting in complex medical surgeries (2. Healthcare: Revolutionizing Patient Care).
   By 2050, AI-powered robotic systems will assist in the majority of complex surgical procedures, improving precision and outcomes.

7. AI in mental health support providing personalized therapy (2. Healthcare: Revolutionizing Patient Care).
   By 2050, AI will provide personalized mental health support, offering real-time counseling and therapy sessions.

8. AI predicting and preventing pandemics (2. Healthcare: Revolutionizing Patient Care).
   By 2050, AI will be used by global health organizations to predict and prevent pandemics, with a 90% accuracy rate in early detection.

9. Fully autonomous vehicles on the roads (2. Automotive: Beyond Autonomous Vehicles).
   By 2050, fully autonomous vehicles will account for at least half of all vehicles on the road, operating without human intervention.

10. AI enhancing personalized shopping experiences (2. Retail: Customizing Consumer Experience).
    By 2050, AI will provide personalized shopping recommendations and experiences for the majority of online shoppers, increasing customer satisfaction and sales.

11. AI-powered dynamic pricing models in retail (2. Retail: Customizing Consumer Experience).
    By 2050, AI-driven dynamic pricing models will be implemented by retail chains, optimizing prices based on real-time demand and inventory.

12. AI systems predicting and preventing equipment failures (2. Manufacturing: Smart Factories).
    By 2050, predictive maintenance systems powered by AI will be used in most manufacturing plants, reducing equipment downtime by at least 50%.

13. AI creating real-time fraud detection systems (2. Financial Sector: Risk, Fraud, and Personalization).
    By 2050, AI-powered fraud detection systems will be used by 90% of financial institutions, significantly reducing fraud incidents.

14. AI in managing personal finances and investments (2. Financial Sector: Risk, Fraud, and Personalization).
    By 2050, AI will manage personal finances, providing real-time budgeting, investment advice, and automated bill payments to optimize financial health.

15. AI automating agricultural processes (2. Agriculture: Precision Farming and Sustainable Practices).
    By 2050, AI-powered automation will handle most agricultural tasks, from planting to harvesting, improving efficiency and reducing labor costs.

16. AI improving agricultural yield predictions (2. Agriculture: Precision Farming and Sustainable Practices).
    By 2050, AI will provide accurate yield predictions for 80% of farming operations, helping to optimize planting and harvesting schedules.

17. AI enhancing creative processes in music and arts (2. Entertainment and Content Creation: Music, Video, Art, Game and Design).
    By 2050, AI will be used in the creation of half of all new music and art, collaborating with human artists to push creative boundaries.

18. AI creating fully immersive virtual reality experiences (2. Entertainment and Content Creation: Music, Video, Art, Game and Design).
By 2050, AI will generate fully immersive virtual reality environments that are indistinguishable from real life, used in the majority of entertainment and training applications.

19. AI transforming real-time sports analytics and performance tracking (2. Entertainment and Content Creation: Music, Video, Art, Game and Design).
By 2050, AI will be integral to sports analytics, with most professional teams using AI to track performance and develop strategies.

20. AI in autonomous military systems (2. Defense: The Future of Warfare).
By 2050, AI-powered autonomous military systems will be deployed in at least half of all military operations, enhancing strategic capabilities and reducing human casualties.

21. AI systems predicting global economic trends with high accuracy (3. Business Innovation: Identifying Opportunities for Growth).
By 2050, AI will accurately predict global economic trends at least one year in advance with an accuracy rate of 90%.

22. AI-powered business innovation platforms (3. Business Innovation: Identifying Opportunities for Growth).
By 2050, over half of Fortune 500 companies will use AI platforms to identify and capitalize on new business opportunities.

23. AI-driven market analysis tools for startups (3. AI Startups: Innovation and Challenges).
By 2050, AI tools will assist new startups in market analysis and business planning, significantly increasing their survival rate.

24. AI in predictive maintenance for business infrastructure (3. Supply Chain Management: A Revolution in Logistics).
By 2050, AI-driven predictive maintenance will be used in the vast majority of business infrastructure, reducing downtime and repair costs.

25. AI in personalized marketing strategies (3. Marketing: Personalization and Consumer Behavior).
    By 2050, AI-driven personalized marketing will be standard practice for businesses, increasing conversion rates and customer loyalty.

26. AI in dynamic pricing models for services (3. Marketing: Personalization and Consumer Behavior).
    By 2050, AI-driven dynamic pricing models will be used by most service-based businesses, optimizing pricing strategies.

27. AI handling the majority of customer service interactions (3. Customer Service Automation: Enhancing Efficiency and Satisfaction).
    By 2050, AI systems will handle 90% of customer service interactions across various industries, providing instant and accurate responses to customer inquiries and issues.

28. AI enhancing employee training and development (3. Human Resources: Recruitment and Management).
    By 2050, companies will use AI to personalize employee training programs, improving skills and productivity.

29. AI improving remote work capabilities (3. AI and the Shift to Remote Work: Long-term Implications).
    By 2050, AI tools will enhance remote work for 90% of companies, enabling seamless collaboration and productivity.

30. Real-time language translation technology widely available (4. Generative AI in Communication).
    By 2050, AI-powered real-time translation will be available on communication devices, allowing seamless conversations across languages.

31. AI-driven personal digital assistants in every household (4. Generative AI in Communication).
    By 2050, households will have AI-driven personal digital assistants that manage daily tasks, schedules, and smart home devices.

32. AI-powered blockchain systems for secure transactions (4. AI and Blockchain: Enhancing Security and Transparency).
    By 2050, AI-integrated blockchain systems will be used in financial transactions, ensuring secure and transparent operations.

33. AI enhancing cybersecurity with adaptive threat detection (4. AI and Blockchain: Enhancing Security and Transparency).
    By 2050, AI will provide adaptive threat detection integrated with blockchain technology, ensuring secure and transparent data transactions.

34. Intelligent Traffic System (4. Integration of AI with IoT: Smart Everything).
    By 2050, AI-enhanced IoT networks will manage intelligent traffic systems, by using real-time data from connected vehicles, traffic lights, and road sensors.

35. AI optimizing energy grid efficiency (4. Renewable Energy: Optimizing Green Technologies).
    By 2050, AI will optimize the efficiency of energy grids, reducing energy waste and ensuring the reliable integration of renewable energy sources such as solar and wind.

36. AI-powered predictive analytics for disaster response (4. AI in Climate Change Mitigation).
    By 2050, AI will predict and manage responses to natural disasters for majority of disaster-prone areas, minimizing damage and saving lives.

37. AI managing autonomous space missions (4. AI in Space Exploration).
    By 2050, AI will autonomously manage 80% of space missions, from planning to execution, including planetary exploration and resource extraction.

38. AI in developing self-repairing materials for construction (4. AI in Space Exploration).

By 2050, AI will design and implement self-repairing materials used in the majority of new space construction projects, reducing maintenance costs and increasing longevity.

39. AI-driven genetic engineering for disease eradication (4. Genetic Engineering).

    By 2050, AI will play a key role in eradicating at least three major genetic diseases through advanced genetic engineering techniques.

40. AI in predictive genomics and personalized medicine (4. Genetic Engineering).

    By 2050, AI will analyze individual genomes to predict health risks and create personalized medicine plans, resulting in a reduction in the incidence of genetic diseases.

41. AI in developing personalized treatment plans based on genetic profiles (4. Genetic Engineering).

    By 2050, AI will be used to create personalized treatment plans based on individual genetic profiles.

42. AI drafting legislation and legal documents (5. Developing Effective AI Legislation).

    By 2050, AI will assist in drafting most new legislation and legal documents, ensuring clarity and reducing human error.

43. AI in judicial systems for case analysis and management (5. Developing Effective AI Legislation).

    By 2050, AI will be used in judicial systems for case analysis, evidence management, and verdict recommendations, improving legal efficiency.

44. AI in crisis management and disaster response coordination (5. Crisis Management).

    By 2050, AI systems will coordinate disaster response efforts in 80% of natural and man-made crises, reducing response times and saving lives.

45. AI-enhanced public transportation systems (6. AI and the Future of Urban Living: Smart Cities and Infrastructure).
By 2050, AI will manage public transportation networks, optimizing routes, reducing congestion, and providing real-time updates to passengers.

46. AI-driven social platforms fostering meaningful connections (6. Social Media and AI: Transforming Engagement and Content Creation).
By 2050, AI-driven social platforms will foster meaningful connections and relationships, enhancing social well-being and reducing loneliness.

47. AI enabling real-time, interactive content creation (6. Social Media and AI: Transforming Engagement and Content Creation).
By 2050, AI will enable social media users to create real-time, interactive content such as live augmented reality experiences, personalized videos, and holographic projections.

48. AI detecting and mitigating misinformation and harmful content creation (6. Social Media and AI: Transforming Engagement and Content Creation).
By 2050, AI systems will detect and mitigate 99% of misinformation and harmful content on social media platforms within seconds of being posted.

49. AI-driven personal health monitors embedded in everyday clothing (6. The Role of AI in Personal Wellbeing).
By 2050, AI-powered health monitors will be integrated into clothing, continuously tracking vital signs and providing health alerts to wearers.

50. AI-driven virtual personal trainers and fitness advisors (6. The Role of AI in Personal Wellbeing).
By 2050, AI will provide virtual personal trainers, creating personalized fitness plans, monitoring progress, and offering real-time guidance and motivation.

51. AI-powered mental health companions for emotional support (6. The Role of AI in Personal Wellbeing).
    By 2050, AI companions will provide emotional support, offering real-time counseling and stress-relief techniques to help individuals manage mental health.

52. AI-powered eldercare and support systems (6. The Role of AI in Personal Wellbeing).
    By 2050, AI will assist in eldercare, providing companionship, monitoring health, and ensuring safety for the elderly population.

53. AI-enhanced personalized learning environments in homes (6. AI and Education: Transforming Learning Environments).
    By 2050, AI will create personalized learning environments for students at home, adapting to their learning styles and providing real-time feedback and support.

54. AI-driven personal hobby and skill development (6. AI and Education: Transforming Learning Environments).
    By 2050, AI will help individuals develop new hobbies and skills, providing personalized lessons, practice schedules, and progress tracking.

55. AI in managing household chores and maintenance (6. AI's Influence on Daily Life by 2050).
    By 2050, AI-powered robots will handle household chores such as cleaning, laundry, and gardening, freeing up time for leisure and family activities.

56. AI-powered smart kitchens for personalized meal planning and preparation (6. AI's Influence on Daily Life by 2050).
    By 2050, AI will manage smart kitchens, suggesting recipes based on dietary needs, and preferences, and even preparing meals with robotic assistance.

57. AI in predictive maintenance for home appliances and infrastructure (6. AI's Influence on Daily Life by 2050).
    By 2050, AI will predict and prevent failures in home appliances and infrastructure, scheduling maintenance and repairs before issues arise.

58. AI-powered universal translators for real-time communication (6. AI's Influence on Daily Life by 2050).
    By 2050, AI devices will provide real-time translation for conversations between people speaking different languages, facilitating global communication and understanding.

59. AI in personalized travel planning and management (6. AI's Influence on Daily Life by 2050).
    By 2050, AI will plan and manage travel itineraries, booking transportation, accommodation, and activities tailored to individual preferences and budgets.

60. AI in enhancing social interactions and community building (6. AI's Influence on Daily Life by 2050).
    By 2050, AI will facilitate social interactions, connecting people with similar interests, organizing events, and fostering a sense of community.

61. AI-powered predictive analytics for investment opportunities (7. Identifying AI Investment Opportunities).
    By 2050, AI will provide real-time predictive analytics to identify the most promising investment opportunities with an accuracy rate of 95%.

62. AI-driven personalized investment portfolios (7. Identifying AI Investment Opportunities).
    By 2050, AI will create and manage personalized investment portfolios for individuals, optimizing returns based on their risk tolerance and financial goals.

63. AI enhancing financial forecasting and economic modeling (7. The Role of AI in Financial Forecasting).
    By 2050, AI will provide advanced financial forecasting and economic modeling, reducing the margin of error in economic predictions to less than 5%.

64. AI in automated stock trading and market making (7. Market Trends and AI's Expanding Influence).
    By 2050, AI will dominate stock trading and market making, executing trades with precision and optimizing market liquidity and stability.

65. AI in cybersecurity, detecting and preventing cyber threats (8. AI and Cybersecurity).
    By 2050, AI will detect and prevent 95% of cyber threats using advanced machine learning algorithms and predictive analytics.

66. AI causing significant job displacement in certain industries (8. AI and Job Displacement).
    By 2050, AI will lead to the displacement of up to half of the jobs in industries such as manufacturing and transportation, necessitating extensive retraining programs.

67. AI enhancing democratic processes through better data analysis (8. AI and Democratic Processes).
    By 2050, AI will be used to analyze public opinion data and improve policy-making processes, enhancing democratic governance in the majority of countries.

68. AI contributing to high energy demand and environmental impact (8. AI and High Energy Demand).
    By 2050, AI data centers will consume 10% of global energy, increasing the carbon footprint and necessitating more sustainable energy solutions.

69. AI improving fairness and reducing bias in decision-making processes (8. The Bias Problem: Making AI Fair for All).
    By 2050, AI algorithms will reduce bias in hiring, lending, and legal decisions, promoting fairer outcomes.

70. AI exacerbating socioeconomic inequalities (8. The Bias Problem: Making AI Fair for All).
By 2050, the unequal distribution of AI technologies will widen the socioeconomic gap, with the top 10% of earners benefiting disproportionately from AI advancements.

71. AI systems experiencing ethical failures and unintended consequences (8. The Bias Problem: Making AI Fair for All).
By 2050, some of AI systems will experience ethical failures or unintended consequences, necessitating robust ethical guidelines and oversight.

72. AI amplifying misinformation and deepfake technologies (8. Blurring the Lines: Is It Still Reality?).
By 2050, AI-generated misinformation and deepfakes will become highly sophisticated, challenging the ability to discern truth from falsehood.

73. AI creating new forms of digital addiction (8. Blurring the Lines: Is It Still Reality?).
By 2050, AI-driven personalized content will lead to increased digital addiction, affecting the mental health.

74. AI causing significant ethical dilemmas in human-robot relationships (8. Blurring the Lines: Is It Still Reality?).
By 2050, AI-powered robots will form relationships with humans, leading to ethical dilemmas about the nature of these interactions and their impact on society.

75. AI developing unexpected behaviors under stress (8. Robot Supervisor Suicide).
By 2050, AI systems in critical roles will need advanced monitoring and safety protocols to prevent unpredictable behaviors or self-destructive actions.

76. AI exacerbating geopolitical tensions and conflicts (8. Stephen Hawking: Existential Risks).
    By 2050, the race for AI supremacy will increase geopolitical tensions, leading to conflicts over AI resources and capabilities.

77. AI enhancing surveillance and privacy invasion (8. Stephen Hawking: Existential Risks).
    By 2050, AI-driven surveillance systems will be used by majority of governments, raising concerns about privacy and civil liberties.

78. AI identifying and bridging skills gaps in the workforce (9. Developing an AI-ready Workforce).
    By 2050, AI will identify skills gaps and provide targeted training programs, reducing unemployment rates in tech-driven industries.

79. AI in forecasting and planning for future job markets (9. Developing an AI-ready Workforce).
    By 2050, AI will accurately forecast future job market trends, helping educational institutions align their curricula with industry needs.

80. AI in global talent matching and remote work platforms (9. Developing an AI-ready Workforce).
    By 2050, AI-powered global talent matching platforms will connect the workforce to remote job opportunities, aligning skills with job requirements across the globe.

81. AI facilitating lifelong learning and skill development (9. Lifelong Learning in an AI World).
    By 2050, 90% of adults will engage in lifelong learning programs powered by AI, enabling them to continuously develop new skills and adapt to changing job markets.

82. AI enhancing the adaptability of educational curricula (9. Lifelong Learning in an AI World).
    By 2050, AI will enable educational institutions to adapt their curricula in real-time, ensuring they remain relevant and up-to-date.

83. AI improving the scalability of educational resources (9. Lifelong Learning in an AI World).
By 2050, AI will enable the scalable delivery of educational resources, reaching 95% of students, including those in remote or underserved areas.

84. AI-driven community engagement platforms for public feedback (9. Addressing Public Perception of AI: Fears and Misconceptions).
By 2050, AI will power community engagement platforms, collecting and analyzing public feedback to improve decision-making in municipalities.

85. AI in fostering public trust through transparent AI systems (9. Addressing Public Perception of AI: Fears and Misconceptions).
By 2050, 90% of AI systems will include transparent algorithms and decision-making processes, fostering public trust and understanding of AI technologies.

86. AI improving accessibility and inclusion for disabled individuals (9. Building Inclusive AI: Ensuring Fairness and Accessibility).
By 2050, AI will create technologies that improve accessibility and inclusion for 90% of disabled individuals, enhancing their quality of life.

87. AI-driven personalized healthcare for all (10. Emergence of New AI-Driven Industries).
By 2050, 90% of people will have access to AI-driven personalized healthcare plans tailored to their genetic profiles and lifestyle.

88. AI-driven personalized education platforms (10. Emergence of New AI-Driven Industries).
By 2050, AI-driven personalized education platforms will be widely adopted in schools, colleges, and universities.

89. AI-driven food production ensuring food security (10. Emergence of New AI-Driven Industries).

    By 2050, AI-driven food production systems will ensure food security for the global population, reducing hunger and malnutrition.

90. AI in developing next-generation quantum communication networks (10. Emergence of New AI-Driven Industries).

    By 2050, AI will develop quantum communication networks that will be adopted by majority of communication infrastructure for secure and instantaneous data transfer.

91. 20-hour work week with maintained productivity (10. The Future Reality of Life: 20 Hour Work Week).

    By 2050, majority of the workforce will enjoy a 20-hour work week while maintaining current levels of productivity through AI-enhanced efficiency.

92. Universal basic income supported by AI-driven economies (10. The Future Reality of Life: 20 Hour Work Week).

    By 2050, majority of countries will implement universal basic income funded by the productivity gains from AI-driven economies, reducing poverty and inequality.

93. BCI enhancing cognitive functions and memory retention (10. Advancements in Brain-Computer Interfaces).

    By 2050, BCI chips designed to enhance cognitive functions and memory retention will improve learning capabilities and cognitive performance.

94. BCI chips enabling seamless human-computer interaction for everyday tasks retention (10. Advancements in Brain-Computer Interfaces).

    By 2050, brain-computer interfaces will perform everyday tasks such as controlling smart home devices, communicating with AI assistants, and managing personal schedules.

95. AI in regenerative medicine and organ growth (10. AI and Lifespan Extension).
By 2050, AI will advance regenerative medicine, enabling the growth of replacement organs from a patient's own cells, reducing the need for organ transplants and increasing the average lifespan by 15 years.

96. Average life expectancy approaching 100 years (10. AI and Lifespan Extension).
By 2050, advances in healthcare and AI will increase the average life expectancy to close to 100 years in most developed countries.

97. Widespread adoption of AI companions for social and emotional support (10. AI and the Evolution of Human-Machine Relationships).
By 2050, AI companions designed to provide social and emotional support will be commonly used.

98. Development of AI systems with generalized problem-solving capabilities (10. Artificial General Intelligence: Approaching Reality).
By 2050, AI systems classified as AGI will be able to perform a wide range of tasks across different domains at a human-equivalent level.

99. AI in creating hyper-realistic virtual simulations (10. Artificial General Intelligence: Approaching Reality).
By 2050, AGI will develop hyper-realistic virtual simulations that are indistinguishable from reality, used for education, training, and entertainment.

100. Self-improving AI systems achieving significant advancements (10. AI Singularity: Getting Closer).
By 2050, AI systems capable of recursive self-improvement will lead to significant advancements in intelligence, approaching human cognitive abilities in many measurable aspects.

APPENDIX 2

Building a Simple AI Project

Welcome to your first AI project! We will build a simple chatbot using Python and natural language processing (NLP) techniques, which we introduced at the beginning. Chatbots are very popular and are an exciting way to learn how AI can understand and respond to human language.

This project will guide you through the key steps of creating a chatbot that can carry on basic conversations. By following these steps, you'll get an idea about the development process and learn how AI processes text, makes decisions and returns relevant responses. By the end, you'll have a basic chatbot that can engage in simple conversations.

Before You Start

1. Install Python: Python is the programming language we'll use for this project. You can download and install from python.org.

2. Install Libraries: Libraries are collections of pre-built code that make it easier to perform certain tasks, such as text processing. To install the libraries we need, run the following commands in your terminal (or bash) - we'll use nltk (Natural Language Toolkit) for processing text, and numpy to handle data.

   ```
   pip install nltk numpy random
   ```

3. Use Jupyter Notebook or Google Colab: These are interactive, web-based environments that allow you to write and run Python code step by step. You can download **Jupyter Notebook** or use **Google Colab**, which runs directly in your browser and doesn't require installation.

Step 1: Prepare the Data

The first step is to prepare the text data that the chatbot will use to generate responses. This involves importing necessary libraries and loading some basic text data.

1. Import Libraries: Here, we're importing the **nltk** library for natural language processing, **numpy** for numerical operations, and **random** to randomize chatbot responses. We also download two necessary datasets, **punkt** (for tokenizing text) and **wordnet** (a lexical database for lemmatization).

```
import nltk, numpy as np, random, string
nltk.download('punkt')
nltk.download('wordnet')
```

2. Load and Preprocess Data: The chatbot needs some pre-loaded text data to work with. We start by opening a text file and reading its contents into Python. We then split the data into individual sentences and words, which the chatbot will analyze.

```
f = open('chatbot_data.txt', 'r',
errors='ignore')
raw_doc = f.read().lower()
sent_tokens = nltk.sent_tokenize(raw_doc)
word_tokens = nltk.word_tokenize(raw_doc)
```

Step 2: Preprocess Text

Before the chatbot can understand the user's input, we need to clean and simplify the text.

1. Tokenize and Lemmatize: **Tokenization** breaks text into individual words, and **lemmatization** reduces each word to its root form (for example, "running" becomes "run"). These steps help the chatbot recognize and process text more accurately.

```
lemmer = nltk.stem.WordNetLemmatizer()
def LemTokens(tokens):
    return [lemmer.lemmatize(token) for token
    in tokens]
```

Step 3: Build Responses

Now we need to create responses for the chatbot. It will check the user's input, compare it to known phrases, and select the most appropriate response. **Cosine similarity** is used to measure how similar the user's message is to sentences in the dataset. This is a mathematical formula that helps the chatbot determine the closest match and generate the best response.

1. Greet Users: The chatbot will recognize common greetings like "hello" or "hi" and respond with messages, such as "Hello, future citizen of 2050!"

```
GREETING_INPUTS = ("hello", "hi", "greetings")
GREETING_RESPONSES = ["Hello, future citizen
of 2050!", "Hi there, ready for the AI
revolution?", "Greetings from the AI-powered
future!"]

def greeting(sentence):
    for word in sentence.split():
        if word.lower() in GREETING_INPUTS:
            return random.
            choice(GREETING_RESPONSES)
```

2. Generate Chatbot Response: To handle non-greeting inputs, the chatbot will compare the user's message with sentences from its data and use a similarity measure to find the closest match. When the chatbot is unsure or confused, it will provide a predefined fallback response.

```
from sklearn.feature_extraction.text import
TfidfVectorizer
from sklearn.metrics.pairwise import
cosine_similarity

def response(user_response):
    robo_response = ‘’
    sent_tokens.append(user_response)
    TfidfVec =
    TfidfVectorizer(tokenizer=LemTokens,
    stop_words=’english’)
    tfidf = TfidfVec.fit_transform(sent_tokens)
    vals = cosine_similarity(tfidf[-1], tfidf)
    idx = vals.argsort()[0][-2]
    flat = vals.flatten()
    flat.sort()
    if flat[-2] == 0:
        robo_response = “Hmm... my AI skills
        must still be developing. Maybe by
        2050 I’ll be smarter!”
    else:
        robo_response = sent_tokens[idx]
    sent_tokens.remove(user_response)
    return robo_response
```

Step 4: Run the Program

The final step is to run the chatbot in a loop, allowing users to engage in a back-and-forth conversation until they choose to exit.

1. Conversation Loop: This loop allows the user to keep interacting with the chatbot. The chatbot will respond to greetings or provide a response based on the input text. The program ends when the user types "bye."

```
flag = True
print("ChatBot: Hello! How can I help?")

while(flag):
   user_response = input().lower()
   if user_response != 'bye':
     if greeting(user_response):
        print("ChatBot:",
        greeting(user_response))
     else:
        print("ChatBot:",
        response(user_response))
   else:
     flag = False
     print("ChatBot: Goodbye! See you in
     2050!")
```

Summary

This project demonstrates core NLP techniques like tokenization and lemmatization to process text. To enhance it, you can integrate machine learning, allowing the chatbot to learn from user interactions by updating its database with user inputs. Over time, this data would help the chatbot adapt and personalize responses, evolving from static rule-based replies to a more intelligent, dynamic system.

REFERENCES

AAAI. (n.d.). AAAI homepage. Retrieved July 31, 2024, from https://aaai.org/

Agarwal, P. (2022, December 31). Emotional ai is no substitute for empathy. Wired. https://www.wired.com/story/artificial-intelligence-empathy/

Ali, M. (2022, August). Clustering in Machine Learning: 5 Essential Clustering Algorithms. Datacamp. https://www.datacamp.com/blog/clustering-in-machine-learning-5-essential-clustering-algorithms

Allyn, B. (2022, March 16). Deepfake video of Zelenskyy could be "tip of the iceberg" in info war, experts warn NPR. https://www.npr.org/2022/03/16/1087062648/deepfake-video-zelenskyy-experts-war-manipulation-ukraine-russia

AltexSoft. (2020, August 20). Credit card fraud detection with machine learning. AltexSoft. https://www.altexsoft.com/blog/credit-card-fraud-detection/

Amazon. (2021, September). Upskilling 2025 Annual Report. Amazon. https://assets.aboutamazon.com/80/a2/7e01e544475aaf15b4f8d5067cc6/amazon-upskilling-report-final.pdf

Awan, A. A. (2023, August). What are Neural Networks? Datacamp. https://www.datacamp.com/blog/what-are-neural-networks

Bajaj, P. (2024, July 17). Reinforcement learning. GeeksforGeeks. https://www.geeksforgeeks.org/what-is-reinforcement-learning/

Barber, G. (2023, November 14). Google deepmind's ai weather forecaster handily beats a global standard. Wired. https://www.wired.com/story/google-deepmind-ai-weather-forecast/

Bovet, A., & Makse, H. A. (2019). Influence of fake news in Twitter during the 2016 US presidential election. Nature Communications, 10(7). https://doi.org/10.1038/s41467-018-07761-2

Brears, R. C. (2024, January 16). Transforming agricultural water management with ai: Optimizing use and enhancing sustainability. Mark and Focus. https://medium.com/mark-and-focus/transforming-agricultural-water-management-with-ai-optimizing-use-and-enhancing-sustainability-c3f19c8406cc

Callahan, M. (2023, February 23). Algorithms were supposed to reduce bias in criminal justice—Do they? The Brink; Boston University. https://www.bu.edu/articles/2023/do-algorithms-reduce-bias-in-criminal-justice/

Cellan-Jones, R. (2014, December 2). Stephen Hawking warns artificial intelligence could end mankind. BBC News. https://www.bbc.com/news/technology-30290540

Cericola, R. (2024, April 4). The best smart assistive devices for people with disabilities. The New York Times. https://www.nytimes.com/wirecutter/reviews/best-assistive-smart-home-technology-for-disabled/

Chappell, B. (2019, November 4). 4-day workweek boosted workers' productivity by 40%, Microsoft Japan says. NPR. https://www.npr.org/2019/11/04/776163853/microsoft-japan-says-4-day-workweek-boosted-workers-productivity-by-40

Chow, R. (2021, September 30). Dartmouth summer research project: The birth of artificial intelligence. History of Data Science. https://www.historyofdatascience.com/dartmouth-summer-research-project-the-birth-of-artificial-intelligence/

Copeland, B. J. (1998). Methods and goals in AI | Artificial intelligence. In Encyclopedia Britannica. Last updated June 28, 2024. https://www.britannica.com/technology/artificial-intelligence/Methods-and-goals-in-AI

Corbyn, Z. (2024, June 29). AI scientist Ray Kurzweil: 'We are going to expand intelligence a millionfold by 2045.' The Guardian. https://www.theguardian.com/technology/article/2024/jun/29/ray-kurzweil-google-ai-the-singularity-is-nearer

Correlation One. (n.d.). How these 5 companies are shaping the workforce of tomorrow. Correlation One. Retrieved July 11, 2024, from https://www.correlation-one.com/blog/how-these-5-companies-are-shaping-the-workforce-of-tomorrow

Daley, S., & Whitfield, B. (2023, February 24). Blockchain & ai: How they integrate & 26 examples. Built In. https://builtin.com/artificial-intelligence/blockchain-ai-examples

Dastin, J. (2018, October 10). Insight—Amazon scraps secret AI recruiting tool that showed bias against women. Reuters. https://www.reuters.com/article/world/insight-amazon-scraps-secret-ai-recruiting-tool-that-showed-bias-against-women-idUSKCN1MK0AG/

Datrics. (n.d.). Ai credit scoring: The future of credit risk assessment. Datrics. Retrieved July 31, 2024, from https://www.datrics.ai/articles/the-essentials-of-ai-based-credit-scoring

Davenport, T. H., & Mittal, N. (2022, November 14). How generative ai is changing creative work. Harvard Business Review. https://hbr.org/2022/11/how-generative-ai-is-changing-creative-work

Digiconomist. (n.d.). Bitcoin energy consumption index. Digiconomist. Retrieved July 31, 2024, from https://digiconomist.net/bitcoin-energy-consumption/

Eightfold AI. (2021, June 1). 4 successful examples of reskilling and upskilling programs. Eightfold AI. https://eightfold.ai/blog/reskilling-and-upskilling/

EPA. (2022, July 25). Epa artificial intelligence inventory [Collections and Lists]. https://www.epa.gov/data/epa-artificial-intelligence-inventory

European Commission. (n.d.). Data protection in the eu—European commission. European Commission. Retrieved July 31, 2024, from https://commission.europa.eu/law/law-topic/data-protection/data-protection-eu_en

Favela, R. (2024, March 12). Weather ai: How artificial intelligence is revolutionizing weather forecasting. Tomorrow.Io. https://www.tomorrow.io/blog/how-weather-ai-is-revolutionizing-weather-forecasting/

Frangoul, A. (2019, December 17). Bosch develops A.I. system in cars to detect distracted, tired drivers. CNBC. https://www.cnbc.com/2019/12/17/bosch-develops-ai-system-in-cars-to-detect-distracted-tired-drivers.html

FutureFit AI. (n.d.). FutureFit ai homepage. Retrieved July 31, 2024, from https://www.futurefit.ai/

GeeksforGeeks. (2024, June 19). Machine learning algorithms. GeeksforGeeks. https://www.geeksforgeeks.org/machine-learning-algorithms/

Giles, M. (2018, February 21). The GANfather: The man who's given machines the gift of imagination. MIT Technology Review. https://www.technologyreview.com/2018/02/21/145289/the-ganfather-the-man-whos-given-machines-the-gift-of-imagination/

Gillis, A. S. (n.d.). Natural language processing (NLP). Enterprise AI | Tech Target. Retrieved July 31, 2024, from https://www.techtarget.com/searchenterpriseai/definition/natural-language-processing-NLP

Goldstein, L. A. (2020, June 1). Speech-to-Text apps for the deaf community. Hearing Like Me. https://www.hearinglikeme.com/speech-to-text-apps-for-the-deaf-community/

Google. (n.d.). What is artificial intelligence (AI) in finance? Google Cloud. Retrieved July 31, 2024, from https://cloud.google.com/discover/finance-ai

Graham, C. (2022, January 26). Robot performs first laparoscopic surgery without human help. Hub. https://hub.jhu.edu/2022/01/26/star-robot-performs-intestinal-surgery/

Gunda, N. S., Gautam, S., & Mitra, S. (2018). Artificial intelligence for water quality monitoring. ECS Meeting Abstracts, MA2018-02(56), 1997. https://doi.org/10.1149/MA2018-02/56/1997

Gupta, P., Chopra, C., & Kasare, A. (2024, May 24). How venture capital is investing in AI in the top five global economies—And shaping the AI ecosystem. World Economic Forum. https://www.weforum.org/

agenda/2024/05/these-5-countries-are-leading-the-global-ai-race-heres-how-theyre-doing-it/

Harrigan Jr., P. (2022, November 17). How amazon uses ai to dominate ecommerce: Top 5 use cases. GoDataFeed. http://www.godatafeed.com/blog/how-amazon-uses-ai-to-dominate-ecommerce

Harris, M. (2022, August 4). Tesla's autopilot depends on a deluge of data. IEEE Spectrum. https://spectrum.ieee.org/tesla-autopilot-data-deluge

Hawking, S. (2016, December 1). This is the most dangerous time for our planet. The Guardian. https://www.theguardian.com/commentisfree/2016/dec/01/stephen-hawking-dangerous-time-planet-inequality

History of Data Science. (2021, September 1). Ai winter: The highs and lows of artificial intelligence. History of Data Science. https://www.historyofdatascience.com/ai-winter-the-highs-and-lows-of-artificial-intelligence/

Horwood, C. (2023, September 27). The role of ai in optimising solar panel efficiency. TechRound. https://techround.co.uk/artificial-intelligence/ai-optimising-solar-panel-efficiency/

Hyper-reach. (2023, August 17). Can ai predict natural disasters? Hyper Reach. https://www.hyper-reach.com/can-ai-predict-natural-disasters/

IBM. (n.d. a). AI Fairness 360. IBM Research Trusted AI. Retrieved July 17, 2024, from https://aif360.res.ibm.com/

IBM. (n.d. b). What is a decision tree? IBM. Retrieved July 31, 2024, from https://www.ibm.com/topics/decision-trees

ICDL. (n.d.). Icdl homepage. Retrieved July 31, 2024, from https://icdl.org/

Interactiva. (n.d.). Interactiva homepage. Retrieved July 31, 2024, from https://interactiva.studio/

Jamrozik, N. (2019, June 25). Senior care: How iot solutions allow the elderly to live independently. IoT For All. https://www.iotforall.com/iot-solutions-that-improve-senior-care

Javaid, S. (2024, March 18). Top 12 supply chain ai use cases in 2024. AIMultiple. https://research.aimultiple.com/supply-chain-ai/

Jee, C. (2018, October 26). A controversial artwork created by AI has hauled in $435,000 at auction. MIT Technology Review. https://www.technologyreview.com/2018/10/26/139292/a-controversial-artwork-created-by-ai-has-hauled-in-435000-at-auction/

jeffbullas.com. (2024, April 11). Check out these 5 crazy examples of ai generated music (Scary good!). Jeffbullas.Com. https://www.jeffbullas.com/ai-generated-music/

Joint Preservation Institute. (n.d.). The Role of AI in Joint Replacement. Joint Preservation Institute. https://www.jointpreservationinstitute.com/post/the-role-of-ai-in-joint-replacement

Kanungo, Y. (2022, February 24). Case study: Amazon ads uses pytorch and aws inferentia to scale models for ads processing. PyTorch. https://pytorch.org/blog/amazon-ads-case-study/

Kissinger, H. A., Schmidt, E., & Huttenlocher, D. (2021). The age of AI: And our human future. Little, Brown and Company.

Klavins, A. (2023, May 13). 10 augmented reality retail examples for better customer experience. Overlyapp. https://overlyapp.com/blog/10-augmented-reality-retail-examples-for-customer-experiences/

Kokoszka, P. (2018, August 21). Chatbots in retail: Nine companies using AI to improve customer experience. Retail Insight Network. https://www.retail-insight-network.com/features/chatbots-in-retail-ai-experience/

Linders, B. (2018, November 20). Building human interfaces with artificial intelligence. InfoQ. https://www.infoq.com/news/2018/11/human-interfaces-ai/

Lytton, C. (2024, February 16). AI hiring tools may be filtering out the best job applicants. BBC. https://www.bbc.com/worklife/article/20240214-ai-recruiting-hiring-software-bias-discrimination

Machine Learning in Plain English. (2023, May 26). Deep learning course — lesson 5: Forward and backward propagation. Medium. https://medium.com/@nerdjock/deep-learning-course-lesson-5-forward-and-backward-propagation-ec8e4e6a8b92

Makaryan, M. (2023, December 21). Ai is transforming military drones. Inkstick. https://inkstickmedia.com/ai-is-transforming-military-drones/

Malone, S. (2022, March 16). Upskilling for ai and machine learning: Lessons from google. Reciprocality. https://reciprocality.org/upskilling-for-ai-and-machine-learning-lessons-from-google/

Manzi, G. (2024, February 13). Revolutionising defence intelligence, surveillance & reconnaissance (Isr) with ai. Karve. https://www.karveinternational.com/insights/revolutionising-defence-intelligence-surveillance-reconnaissance-isr-with-ai

Marr, B. (2016, December 29). 4 mind-blowing ways facebook uses artificial intelligence. Forbes. https://www.forbes.com/sites/bernardmarr/2016/12/29/4-amazing-ways-facebook-uses-deep-learning-to-learn-everything-about-you/

Marr, B. (2023, April 10). Artificial intelligence in space: The amazing ways machine learning is helping to unravel the mysteries of the universe. Forbes. https://www.forbes.com/sites/bernardmarr/2023/04/10/artificial-intelligence-in-space-the-amazing-ways-machine-learning-is-helping-to-unravel-the-mysteries-of-the-universe/

Martinis, J., & Boixo, S. (2019, October 23). Quantum supremacy using a programmable superconducting processor. Google Research. http://research.google/blog/quantum-supremacy-using-a-programmable-superconducting-processor/

Masood, A., & Ahmad, K. (2021). A review on emerging artificial intelligence (Ai) techniques for air pollution forecasting: Fundamentals, application and performance. Journal of Cleaner Production, 322, 129072. https://doi.org/10.1016/j.jclepro.2021.129072

McGonagle, J., Shaikouski, G., Williams, C., Hsu, A., Khim, J., & Miller, A. (n.d.). Backpropagation. Brilliant. Retrieved July 11, 2024, from https://brilliant.org/wiki/backpropagation/

MITRE. (2024, March 25). MITRE Opens New AI Assurance and Discovery Lab. MITRE. https://www.mitre.org/news-insights/news-release/mitre-opens-new-ai-assurance-and-discovery-lab

Morgan Stanley. (2023, September 15). Ai and cybersecurity: A new era. Morgan Stanley. https://www.morganstanley.com/articles/ai-cybersecurity-new-era

Mount Sinai. (2022, September 20). Mount sinai researchers use artificial intelligence to uncover the cellular origins of alzheimer's disease and other cognitive disorders. Mount Sinai Health System. https://www.mountsinai.org/about/newsroom/2022/mount-sinai-researchers-use-artificial-intelligence-to-uncover-the-cellular-origins-of-alzheimers-disease-and-other-cognitive-disorders

Nanalyze. (2019, October 19). 8 Diagnostics and Imaging Startups for Digital Health. Nanalyze. https://www.nanalyze.com/2019/10/ai-diagnostics-imaging-startups/

NASA. (n.d.). Robonaut. Retrieved August 19, 2024, from https://www.nasa.gov/robonaut2/

Nelson, A. (n.d.). Innovative teaching and learning center: Artificial intelligence introduction: staff resource: weak vs. Strong. Retrieved July 31, 2024, from https://itlc.northwoodtech.edu/ai/introduction/staff/wvs

Neumetric. (n.d.). Gdpr vs pdpa: Understanding the differences. Neumetric. Retrieved July 31, 2024, from https://www.neumetric.com/gdpr-vs-pdpa/

Newberry, C. (2024, March 4). 2024 social media algorithms: A guide for all networks. Social Media Marketing & Management Dashboard. https://blog.hootsuite.com/social-media-algorithm/

NPR Staff. (2016, July 24). The reason your feed became an echo chamber—And what to do about it. NPR. https://www.npr.org/

sections/alltechconsidered/2016/07/24/486941582/the-reason-your-feed-became-an-echo-chamber-and-what-to-do-about-it

NYU Langone Health. (2021, September 24). Artificial intelligence tool improves accuracy of breast cancer imaging. NYU Langone Health. https://nyulangone.org/news/artificial-intelligence-tool-improves-accuracy-breast-cancer-imaging

OpenArt. (n.d.). OpenArt home page. Retrieved July 31, 2024, from https://openart.ai/

Padwick, C. (2020, August 6). AI for agriculture: How PyTorch enables Blue River's robots. The Robot Report. https://www.therobotreport.com/ai-for-agriculture-how-pytorch-enables-blue-rivers-robots/

Pal, K. (2024, January 18). How Can an AI Model Create Music? Techopedia. https://www.techopedia.com/how-can-an-ai-model-create-music

Patel, S. (2024, June 14). Top 10 startups developing ai for internet of things. AI Startups. https://www.ai-startups.org/top/iot/

Paths to Literacy. (n.d.). Seeing ai: Free app narrating world around you. Paths to Literacy. Retrieved July 11, 2024, from https://www.pathstoliteracy.org/resource/seeing-ai-free-app-narrating-world-around-you/

Pegasus One. (n.d.). How Chat GPT utilizes the advancements in Artificial Intelligence to create a revolutionary language model. Pegasus One. Retrieved July 31, 2024, from https://www.pegasusone.com/how-chat-gpt-utilizes-the-advancements-in-artificial-intelligence-to-create-a-revolutionary-language-model/

Raghavan, P. (2020, October 15). How AI is powering a more helpful Google. Google. https://blog.google/products/search/search-on/

RealSpace. (2023, September 22). Advancements in ai for accurate real-time property valuations. RealSpace. https://www.realspace3d.com/blog/real-time-property-valuations-how-ai-algorithms-are-making-it-possible/

Reynolds, M. (2024, January 22). Could generative AI help break down language barriers plaguing the Indian justice system? ABA Journal. https://www.abajournal.com/web/article/could-generative-ai-help-break-down-language-barriers-plaguing-the-indian-justice-system

Rosic, A. (2023, November 23). Proof of work vs proof of stake: Basic mining guide. Blockgeeks. https://blockgeeks.com/guides/proof-of-work-vs-proof-of-stake/

Rusnak, P. (2024, July 23). Can ai tech help your farm stand strong in a storm? Yes. Growing Produce. https://www.growingproduce.com/vegetables/can-ai-tech-help-your-farm-stand-strong-in-a-storm-yes/

Ryan-Mosley, T. (2023, February 20). How AI can actually be helpful in disaster response. MIT Technology Review. https://www.technologyreview.com/2023/02/20/1068824/ai-actually-helpful-disaster-response-turkey-syria-earthquake/

Sec1. (n.d.). The rise of artificial intelligence in cybersecurity. Sec1. Retrieved July 11, 2024, from https://sec1.io/blog/the-rise-of-artificial-intelligence-in-cybersecurity/

Seth, S. (2023, December 14). Basics of algorithmic trading: Concepts and examples. Investopedia. https://www.investopedia.com/articles/active-trading/101014/basics-algorithmic-trading-concepts-and-examples.asp

Slait.ai. (n.d.). Slait – real-time sign language translator with ai. Slait.Ai. Retrieved July 11, 2024, from https://slait.ai

Smith, C. S. (2023, December 11). Top 10 quantum computing companies making change. Forbes. https://www.forbes.com/sites/technology/article/top-quantum-computing-companies/

Stanford Basic Income Lab. (n.d.). What is ubi. The Stanford Basic Income Lab. Retrieved July 31, 2024, from https://basicincome.stanford.edu/about/what-is-ubi/

Startup Genome. (2024). Global Startup Ecosystem Ranking 2024 (Top 40). Startup Genome. https://startupgenome.com/article/global-startup-ecosystem-ranking-2024-top-40

Stofan, D. (2023, January 21). 6 Smart Cities that get traffic control right. Good Vision. https://blog.goodvisionlive.com/6-smart-cities-that-get-traffic-control-right

Stone, P., Brooks, R., Brynjolfsson, E., Ryan, C., Etzioni, O., Greg, H., Hirschberg, J., Kalyanakrishnan, S., Kamar, E., Kraus, S., Leyton-Brown, K., Parkes, D., Press, W., Saxenian, A., Shah, J., Tambe, M., & Teller, A. (2016). Artificial Intelligence and Life in 2030 (One Hundred Year Study on Artificial Intelligence: Report of the 2015-2016 Study Panel). Stanford University. https://ai100.stanford.edu/2016-report

Synthesia. (n.d.). AI video editor. Synthesia. Retrieved July 31, 2024, from https://www.synthesia.io/tools/ai-video-editor

Tastewise. (n.d.). GenAI-Powered Consumer Data Platform for modern food & beverage brands. Tastewise. Retrieved July 15, 2024, from https://tastewise.io

Tegmark, M. (2017). Life 3.0: Being human in the age of artificial intelligence. Alfred A. Knopf.

TensorFlow. (n.d.). Case studies and mentions. TensorFlow. Retrieved July 17, 2024, from https://www.tensorflow.org/about/case-studies

Teo, P. (2022, February 21). Healthcare virtual assistants: Use cases, examples & benefits. Key Reply. https://www.keyreply.com/blog/healthcare-virtual-assistants-use-cases

The Editors of Encyclopedia Britannica. (1998, July 20). Turing test | artificial intelligence. Encyclopedia Britannica; Last updated May 31, 2024. https://www.britannica.com/technology/Turing-test

Tierney, J. (2008, August 26). Technology that outthinks us: A partner or a master? The New York Times. https://www.nytimes.com/2008/08/26/science/26tier.html

Tiffany. (2023, November 29). Top 3 real-life retail chatbot examples. ChatInsight. https://www.chatinsight.ai/chatbots/retail-chatbot-examples/

USPTO. (2024, February 12). USPTO issues inventorship guidance and examples for AI-assisted inventions. USPTO. https://www.uspto.gov/subscription-center/2024/uspto-issues-inventorship-guidance-and-examples-ai-assisted-inventions

Vincent, J. (2024, February 16). How much electricity does AI consume? The Verge. https://www.theverge.com/24066646/ai-electricity-energy-watts-generative-consumption

Whooley, S. (2022, July 19). Synchron completes its first U.S. human implant of brain-computer interface rivaling Elon Musk's Neuralink. MassDevice. https://www.massdevice.com/synchron-brain-computer-interface-elon-musk-neuralink/

Williams, J. (2024, April 2). The rise of virtual influencers: Are they here to stay? Sprout Social. https://sproutsocial.com/insights/virtual-influencers/

Wodecki, B. (2022, July 12). AI to Help NASA's James Webb Telescope Map the Stars. AI Business. https://aibusiness.com/verticals/ai-to-help-nasa-s-james-webb-telescope-map-the-stars

Zhu, J., & Isaacs, R. (2024, May 29). Campaign microtargeting and AI can jeopardize democracy. British Politics and Policy at LSE. https://blogs.lse.ac.uk/politicsandpolicy/campaign-microtargeting-and-ai-can-jeopardize-democracy/

THE AUTHOR

MARCIN TKACZYK

SAP architect, developer, consultant, and director Marcin Tkaczyk is the author of *The AI Revolution 2050: How Artificial Intelligence Will Transform Health, Economy and Daily Life by Mid-Century.*

Born in Poland, he has over two decades' worth of experience in the global IT business sector, where he specializes in SAP solutions for enterprise resource planning. Marcin has traveled and worked in many countries, witnessing the exponential growth of the IT industry. He holds an MA in Economics and Financial Markets, obtained from the SGH Warsaw School of Economics.

His first book delves into the subject of AI, as artificial intelligence has been one of the most exciting discoveries of his explorations. Currently focused on generative AI, Marcin is also studying other subsets, including NLP, ML, and Computer vision, as well as underlying systems (such as smart cities, smart homes, and robotics). He actively participates in the global AI community and AI projects, striving to incorporate solutions into the real world and demonstrate his deep commitment to the field.

In his free time, Marcin enjoys traveling to discover new places and photography. He is enthusiastic about nutritional science and tries to adhere to a healthy, balanced lifestyle and practice weight and personal training. He is also passionate about running, skiing, cinema, literature, and kitesurfing (a skill he admits he has yet to hone)!

Leave a

REVIEW

Made in United States
North Haven, CT
10 November 2026

22096544R00120